Field Techniques for Estimating Water Fluxes Between Surface Water and Ground Water

Edited by Donald O. Rosenberry and James W. LaBaugh

Techniques and Methods 4–D2

U.S. Department of the Interior
U.S. Geological Survey

U.S. Department of the Interior
DIRK KEMPTHORNE, Secretary

U.S. Geological Survey
Mark D. Myers, Director

U.S. Geological Survey, Reston, Virginia: 2008

For product and ordering information:
World Wide Web: http://www.usgs.gov/pubprod
Telephone: 1-888-ASK-USGS

For more information on the USGS—the Federal source for science about the Earth, its natural and living resources, natural hazards, and the environment:
World Wide Web: http://www.usgs.gov
Telephone: 1-888-ASK-USGS

Suggested citation:
Rosenberry, D.O., and LaBaugh, J.W., 2008, Field techniques for estimating water fluxes between surface water and ground water: U.S. Geological Survey Techniques and Methods 4–D2, 128 p.

Contents

Conversion Factors, Definitions, and Abbreviations

Inch/Pound to SI

Multiply	By	To obtain
Length		
kilometer (km)	0.6214	mile (mi)
meter (m)	3.281	foot (ft)
centimeter (cm)	0.3937	inch (in)
millimeter (mm)	0.03937	inch (in)
Area		
square kilometer (km^2)	0.3861	square mile (mi^2)
square meter (m^2)	10.76	square foot (ft^2)
square centimeter (cm^2)	0.1550	square inch (in^2)
square centimeter (cm^2)	0.001076	square foot (ft^2)
Volume		
liter (L)	0.2642	gallon (gal)
liter (L)	1.057	quart (qt)
liter (L)	61.02	cubic inch (in^3)
milliliter (mL)	0.06102	cubic inch (in^3)
cubic centimeter (cm^3)	0.06102	cubic inch (in^3)
cubic meter (m^3)	264.2	gallon (gal)
cubic meter (m^3)	35.31	cubic foot (ft^3)
Flow rate, velocity		
milliliter per minute (mL/min)	0.06102	cubic inch per minute (in^3/min)
liter per minute (L/min)	0.2642	gallons per minute (gpm)
centimeter per day (cm/d)	0.0328	feet per day (ft/d)
meter per second (m/s)	283461	feet per day (ft/d)
Pressure		
kilopascal (kPa)	0.1450	pound per square inch (psi)
kilopascal (kPa)	0.009869	atmosphere, standard (atm)
kilopascal (kPa)	0.3346	feet of water (at 39 degrees F)
kilopascal (kPa)	0.01	bar
Mass to weight force		
gram (g)	0.0353	ounce (oz)
gram (g)	0.002205	pound (lb)
Velocity		
knot (kn)	1.151	miles per hour (mph)
meter per second (m/s)	2.237	miles per hour (mph)
Discharge		
cubic meters per second (m^3/s)	35.315	cubic feet per second (cfs)
liters per second (L/s)	0.03531	cubic feet per second (cfs)
Thermal conductivity		
Watt per meter per degree Celsius (W/m/°C)	0.5778	BTU per foot-hour per degree Fahrenheit (BTU/ft-hr/°F)
Energy		
Joule (J)	0.0009478	British thermal unit (BTU)

Temperature in degrees Celsius (°C) may be converted to degrees Fahrenheit (°F) as follows:

$$°F = (1.8 \times °C) + 32$$

Temperature in degrees Fahrenheit (°F) may be converted to degrees Celsius (°C) as follows:

$$°C = (°F - 32)/1.8$$

Introduction and Characteristics of Flow

By James W. LaBaugh and Donald O. Rosenberry

Chapter 1 of
**Field Techniques for Estimating Water Fluxes Between
Surface Water and Ground Water**

Edited by Donald O. Rosenberry and James W. LaBaugh

Techniques and Methods Chapter 4–D2

U.S. Department of the Interior
U.S. Geological Survey

Contents

Figures

Chapter 1
Introduction and Characteristics of Flow

By James W. LaBaugh and Donald O. Rosenberry

Introduction

Interest in the use and development of our Nation's surface- and ground-water resources has increased significantly during the past 50 years (Alley and others, 1999; Hutson and others, 2004). At the same time, a variety of techniques and methods have been developed to examine and monitor these water resources. Quantifying the connection between surface water and ground water also has become more important because the use of one of these resources can have unintended consequences on the other (Committee on Hydrologic Science, National Research Council, 2004). In an attempt to convey the importance of the linkages and interfaces between surface water and ground water, the two have been described as a "single resource" (Winter and others, 1998). An improved understanding of the connection between surface and ground waters increasingly is viewed as a prerequisite to effectively managing these resources (Sophocleous, 2002). Thus, water-resource managers have begun to incorporate management strategies that require quantifying flow between surface water and ground water (Danskin, 1998; Bouwer and Maddock, 1997; Dokulil and others, 2000; Owen-Joyce and others, 2000; Barlow and Dickerman, 2001; Jacobs and Holway, 2004).

The use of surface water (or ground water) can change the location, rate, and direction of flow between surface water and ground water (Stromberg and others, 1996; Glennon, 2002; Galloway and others, 2003). Pumping wells in the vicinity of rivers commonly cause river water to flow into the underlying ground-water body, which can affect the quality of the ground water (Childress and others, 1991; McCarthy and others, 1992; Lindgren and Landon, 1999; Steele and Verstraeten, 1999; Zarriello and Reis, 2000; Sheets and others, 2002). In some cases, ground water is pumped to provide water for cooling industrial equipment and then discharged into lakes, ponds, or rivers (Andrews and Anderson, 1978; Hutson and others, 2004). Ground water also may be pumped specifically to maintain lake levels for recreation purposes, especially during droughts (Stewart and Hughes, 1974; Mcleod, 1980; Belanger and Kirkner, 1994; Metz and Sacks, 2002). Surface water can be directed into surface basins where water percolates to the underlying aquifer—a process known as artificial recharge (Galloway and others, 2003). Ground-water discharge areas, where ground water flows into surface water, can be important habitats for fish (Garrett and others, 1998; Power and others, 1999; Malcolm and others, 2003a, 2003b). Water in irrigation canals can flow or seep to an underlying aquifer, which eventually discharges water to rivers, thereby sustaining streamflow essential for the maintenance of fish populations (Konrad and others, 2003).

Interest in the interaction of surface water and ground water is not confined to inland waters. This interaction has been studied in coastal areas because fresh ground-water supplies can be affected by intrusion of saltwater (Barlow and Wild, 2002). Beyond the issue of water supply for human consumption, increased attention has been given to the ground water that discharges to oceans and estuaries, both in terms of water quantity and quality (Bokuniewicz, 1980; Moore, 1996, 1999; Linderfelt and Turner, 2001). Discharge of fresh ground water to oceans and estuaries, also referred to as submarine ground-water discharge, is important in maintaining the flora and fauna that have evolved to exploit this source of fresh water in a saline environment (Johannes, 1980; Simmons, 1992; Corbett and others, 1999). Nitrate in submarine ground-water discharge to estuaries and coastal waters can result in eutrophication of those waters (Johannes, 1980; Johannes and Hearn, 1985; Valiela and others, 1990; Taniguchi and others, 2002). Withdrawals or pumping of ground water at near-shore, inland locations can reduce the submarine discharge of ground water offshore and change the environmental conditions of these settings (Simmons, 1992). Some coral reefs may be endangered by diminished submarine ground-water discharge (Bacchus, 2001, 2002).

The variety of settings of interest for the examination of the interaction between surface water and ground water makes evident the need for methods to describe and quantify that flow. The exact method chosen for each setting will vary depending on the physical and hydrological conditions present in those settings, as well as the scale of the interaction. Some degree of measurement uncertainty accompanies each method or technique. Thus, it is prudent to consider using more than one method to examine the interaction between surface water and ground water. Because numerous techniques and methods are available to describe and quantify the flow between surface water and ground water, it is useful to provide water-resource investigators an overview of available techniques and methods, as well as their application.

Purpose and Scope

Several methods have been developed and applied to the study of the exchange between surface water and ground water (fig. 1). Different methods are better suited for characterizing or measuring flow over large or small areas. If an initial view of a considerable area or distance is needed to determine where measurable ground-water discharge is occurring, aerial infrared photography or imagery can be effective reconnaissance tools. On a smaller scale, some methods may involve direct measurement of sediment temperature or specific conductance along transects within a surface-water body, or use of dyes or other tracers to indicate the direction and rate of water movement. The measurement of water levels in well networks in the watershed can be used to determine ground-water gradients relative to adjacent surface water, which in turn can indicate the direction and rate of flow between the surface-water body and the underlying aquifer. In streams and rivers, measurement of flow at the endpoints of a channel reach can reveal if the reach is gaining flow from ground water or losing flow to ground water. Addition of tracers to streams also can be used to determine surface-water interaction with ground water over a range of scales. Local interaction of surface water with ground

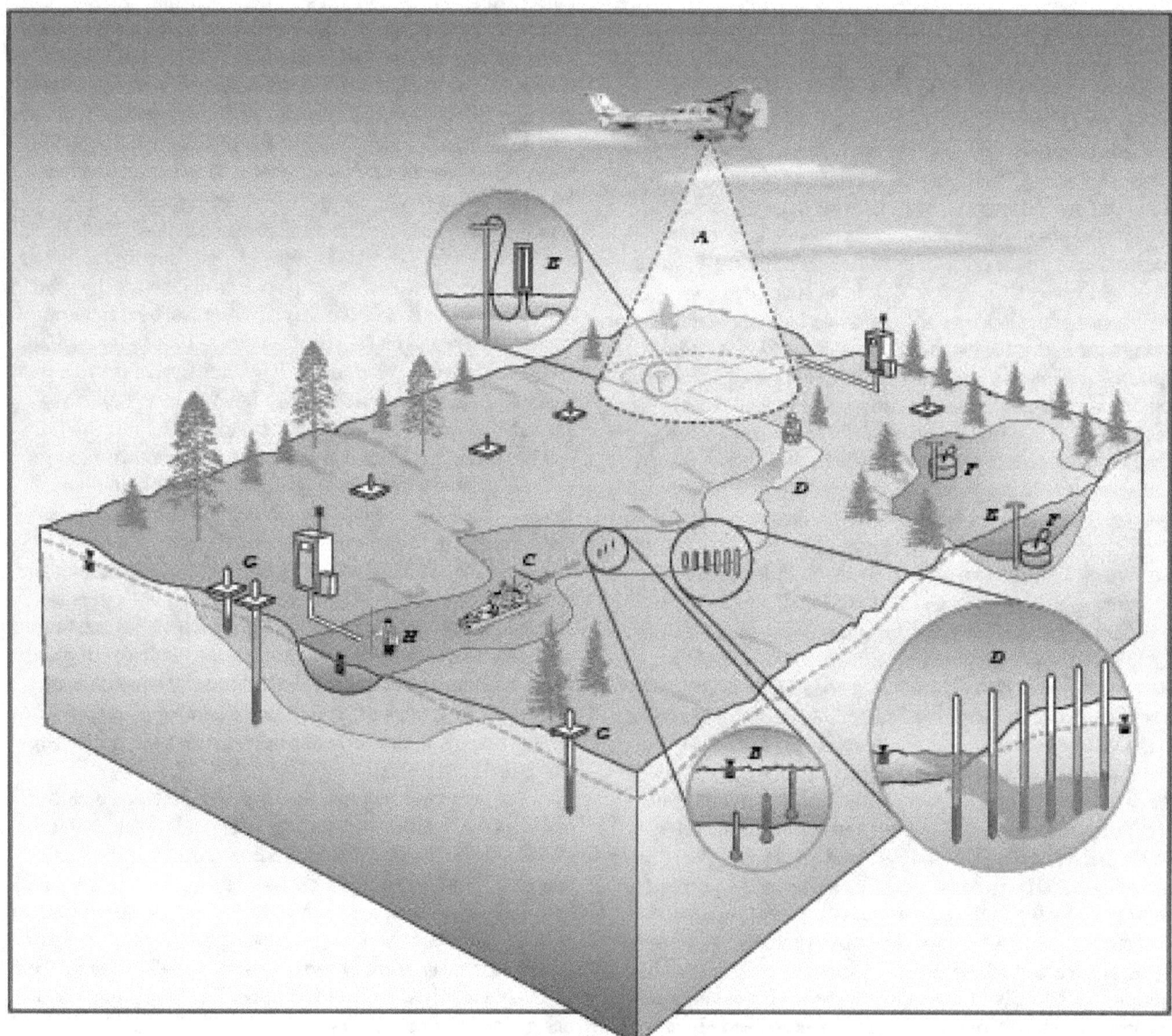

Figure 1.   Summary of techniques that have been used for the measurement or estimation of water fluxes between surface water and ground water. Techniques illustrated include: (*A*) aerial infrared photography and imagery, (*B*) thermal profiling, (*C*) the use of temperature and specific-conductance probes, (*D*) dyes and tracers, (*E*) hydraulic potentiomanometers, (*F*) seepage meters, (*G*) well networks, and (*H*) streamflow measurements. (Artwork by John M. Evans, U.S. Geological Survey, retired.)

water is measured by placing devices such as thermistors, minipiezometers, and seepage meters in the sediment, to monitor temperature gradients, hydraulic gradients, or quantity of flow. Determination of how interaction of surface water with ground water changes over time is made possible by using data-recording devices ("data loggers") in conjunction with pressure transducers, thermistors, and water-quality probes.

This report is designed to make the reader aware of the breadth of approaches (fig. 1) available for the study of the exchange between surface and ground water. To accomplish this, the report is divided into four chapters. Chapter 1 describes many well-documented approaches for defining the flow between surface and ground waters. Subsequent chapters provide an in-depth presentation of particular methods. Chapter 2 focuses on three of the most commonly used methods to either calculate or directly measure flow of water between surface-water bodies and the ground-water domain: (1) measurement of water levels in well networks in combination with measurement of water level in nearby surface water to determine water-level gradients and flow; (2) use of portable piezometers (wells) or hydraulic potentiomanometers to measure hydraulic gradients; and (3) use of seepage meters to measure flow directly. Chapter 3 focuses on describing the techniques involved in conducting water-tracer tests using fluorescent dyes, a method commonly used in the hydrogeologic investigation and characterization of karst aquifers, and in the study of water fluxes in karst terranes. Chapter 4 focuses on heat as a tracer in hydrological investigations of the near-surface environment.

This report focuses on measuring the flow of water across the interface between surface water and ground water, rather than the hydrogeological or geochemical processes that occur at or near this interface. The methods, however, that use hydrogeological and geochemical evidence to quantify water fluxes are described herein. This material is presented as a guide for those who have to examine the interaction of surface water and ground water. The intent here is that both the overview of the many available methods and the in-depth presentation of specific methods will enable the reader to choose those study approaches that will best meet the requirements of the environments and processes they are investigating, as well as to recognize the merits of using more than one approach. To that end, at this point it is useful to examine the content of each chapter in more detail.

Chapter 1 provides an overview of typical settings in the landscape where interactions between surface water and ground water occur. The chapter reviews the literature, particularly recent publications, and describes many well-documented methods for defining the flow between surface and ground waters. A brief overview of the theory behind each method is provided. Information is presented about the field settings where the method has been applied successfully, and, where possible, generalizes the requirements of the physical setting necessary to the success of the method. Strengths and weaknesses of each method are noted, as appropriate. This will aid the investigator in choosing methods to apply to their

setting. For those already familiar with some of these methods, the review of recent literature provides information about improvements in these methods.

Chapter 2 describes three of the most commonly used methods to either calculate or directly measure flow of water between surface-water bodies and the ground-water domain. The first method involves measurement of water levels in a network of wells in combination with measurement of the stage of the surface-water body to calculate gradients and then water flow. The second method involves the use of portable piezometers (wells) or hydraulic potentiomanometers to measure gradients. In the third method, seepage meters are used to directly measure flow across the sediment-water interface at the bottom of the surface-water body. Factors that affect measurement scale, accuracy, sources of error in using each of the methods, common problems and mistakes in applying the methods, and conditions under which each method is well- or ill-suited also are described.

Chapter 3 presents an overview of methods that are commonly used in the hydrogeologic investigation and characterization of karst aquifers and in the study of water fluxes in karst terranes. Special emphasis is given to describing the techniques involved in conducting water-tracer tests using fluorescent dyes. Dye-tracer test procedures described herein represent commonly accepted practices derived from a variety of published and previously unpublished sources. Methods that are commonly applied to the analysis of karst spring discharge (both flow and water chemistry) also are reviewed and summarized.

Chapter 4 reviews early work addressing heat as a tracer in hydrological investigations of the near-surface environment, describes recent advances in the field, and presents selected new results designed to identify the broad application of heat as a tracer to investigate surface-water/ground-water exchanges. An overview of field techniques for estimating water fluxes between surface water and ground water with heat is provided.

To familiarize readers with flow conditions that may occur during their studies, the next section of Chapter 1 describes commonly observed interactions between surface water and ground water.

Characteristics of Water Exchange Between Surface Water and Ground Water

Most measurements made for the purpose of quantifying exchange between surface water and ground water are obtained at points within a short distance of the shoreline of the surface-water body. Shorelines represent the horizontal interface between ground water and surface water, an interface that is highly dynamic spatially and temporally. Because of the complex physical processes that occur in precisely the area where measurements are needed, it is important to understand those processes at shorelines and the range of potential changes in conditions at shorelines that occur over time. The following section elaborates these points.

Typically, a significant break in slope in the water table occurs where the horizontal surface of a lake, stream, or wetland intersects the sloping surface of the ground-water table (fig. 2). Because of this break in slope, ground-water flow lines diverge where they extend beneath and end at the sediment-water interface. Diverging flow lines indicate that the rate of flow per unit area is decreasing. Given homogeneous and isotropic conditions in the porous media adjacent to and beneath the sediment-water interface, seepage across the interface will decrease exponentially with distance from shore (fig. 3) (McBride and Pfannkuch, 1975; Pfannkuch and Winter, 1984). The movement of water between surface water and ground water can occur in a variety of settings or landscapes (fig. 4), each of which can be related to the break in slope of the water table defined by "an upland adjacent to a lowland separated by an intervening steeper slope" (Winter, 2001).

Ground-water flow lines bend substantially beneath the sediment-water interface just before they intersect the surface-water body. Measurements of hydraulic-head gradients typically assume that the flow lines either are horizontal (in the case of comparing heads in near-shore wells with surface-water stage) or vertical (in the case of inserting the screened intervals of wells to some depth beneath the sediment-water interface). In reality, the orientation of the flow lines are somewhere between horizontal and vertical as shown in figure 5.

Characteristics of Near-Shore Sediments

Although some investigators have found that seepage decreases exponentially with distance from shore (Lee, 1977; Fellows and Brezonik, 1980; Erickson, 1981; Attanayake and Waller, 1988; Rosenberry, 1990), other studies report that the decrease in flow across the sediment-water interface is not

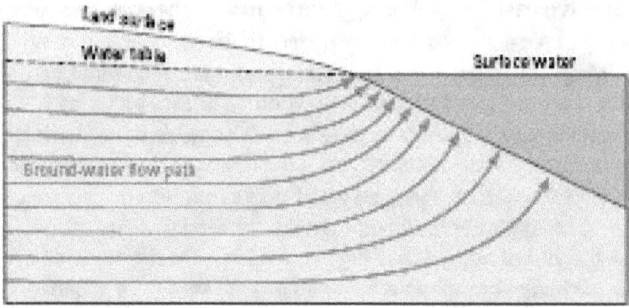

Figure 3.   Decrease in seepage discharge with distance from shore (from Winter and others, 1998).

exponential because of heterogeneity of the sediment. One of the early findings of a departure from what would be expected in a homogeneous, isotropic setting was reported by Woessner and Sullivan (1984) in their study of Lake Mead, Nevada. At many of the transects across which they collected data in Lake Mead, they found seepage did not decrease exponentially, and furthermore, that seepage sometimes decreased and then increased with distance from shore. They reported a large variability in seepage with distance from shore. This variability was attributed to heterogeneity in the sediments in the vicinity of the sediment-water interface. Krabbenhoft and Anderson (1986) also reported that seepage was focused in a gravel lens that intersected the lakebed some distance from shore at Trout Lake, Wisconsin. It now generally is recognized that aquifers adjacent to and beneath surface-water bodies rarely can be considered homogeneous, and usually are not isotropic.

Many processes act to create heterogeneity at the sediment-water interface. A few are listed below.

1. *Fluvial processes*—Depositional and erosional processes occur nearly constantly in streambeds and riverbeds, making heterogeneity a significant feature in these sediment-water interfaces. Organic deposits commonly are buried by deposition of inorganic material, resulting in interlayering of these different sediment types. Channel aggradation and flood scour can cause a shoreline to shift laterally many meters. Seasonal erosion and deposition related to spring floods also create a temporal component to the heterogeneity.

2. *Edge effects*—Shoreline erosion and deposition related to wave action in lakes, large wetlands, and rivers create heterogeneity at the sediment-water interface. Waves erode banks, which subsequently fail as new material slumps into the surface-water body. Fine-grained sediments are moved away from shore, often leaving a cobble- to boulder-sized pavement at the shoreline. Sediment deposition by overland flow commonly results in near-shore, fan-shaped deposits following heavy rainfall. Waves also rework sediments following slump events or sediment transport associated with overland flow, causing movement of fine-grained materials into voids created by movement of cobble- to boulder-sized sediments. In addition, changing surface-water stage causes the position

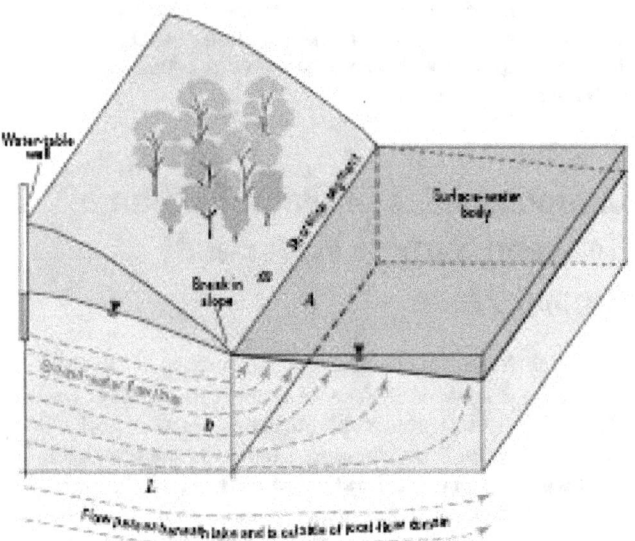

Figure 2.   Typical hydraulic conditions in the vicinity of the shoreline of a surface-water body. (Artwork by Donald O. Rosenberry, U.S. Geological Survey.)

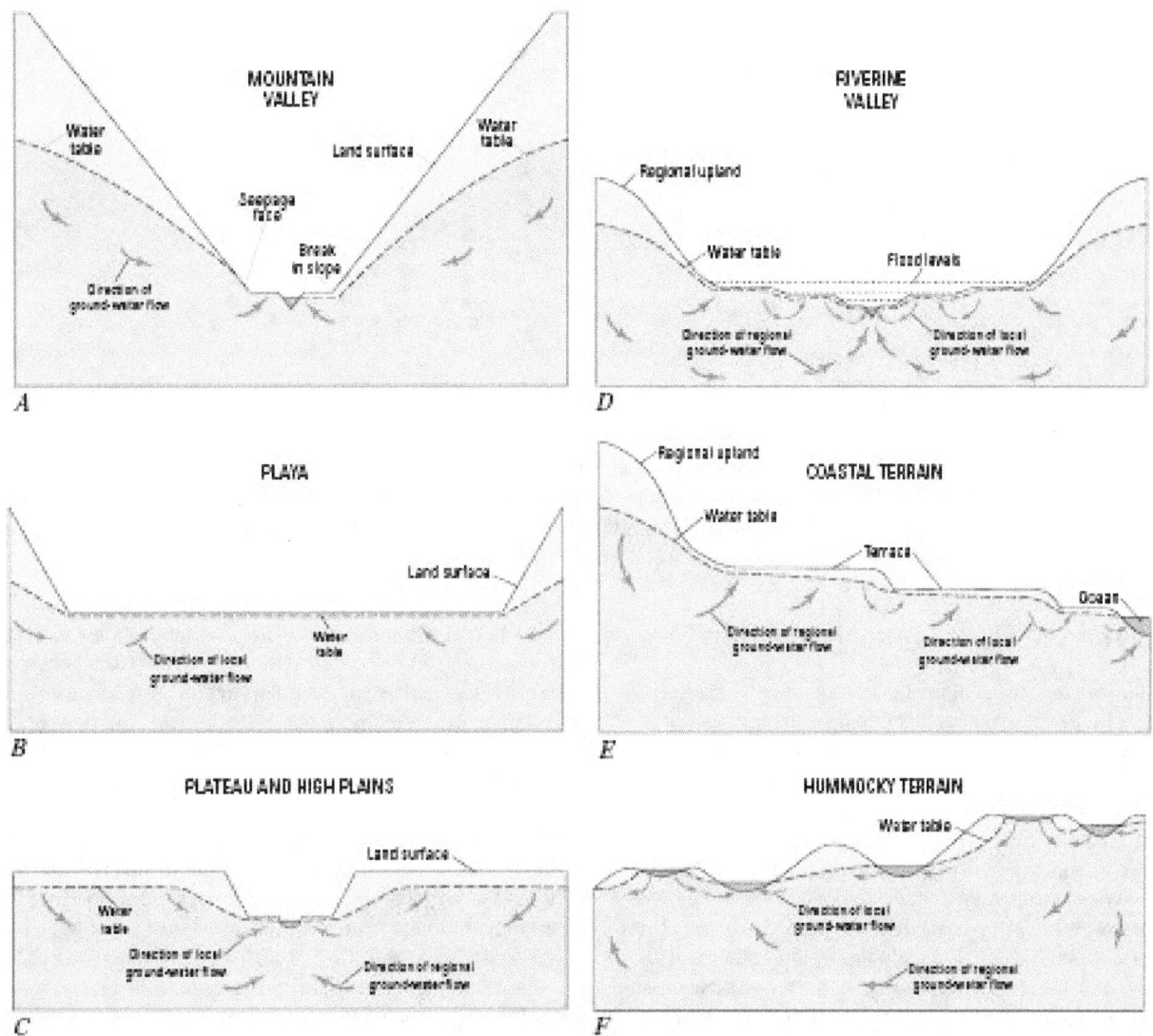

Figure 4.   Generalized hydrologic landscapes: *A*, narrow uplands and lowlands separated by a large steep valley side (mountainous terrain); *B*, large broad lowland separated from narrow uplands by steeper valley sides (playas and basins of interior drainage); *C*, small narrow lowlands separated from large broad uplands by steeper valley side (plateaus and high plains); *D*, small fundamental hydrologic landscape units nested within a large fundamental hydrologic landscape unit (large riverine valley with terraces); *E*, small fundamental hydrologic landscape units superimposed on a larger fundamental hydrologic landscape unit (coastal plain with terraces and scarps); *F*, small fundamental hydrologic landscape units superimposed at random on large fundamental hydrologic landscape units (hummocky glacial and dune terrain) (from Winter, 2001, copyright the American Water Resources Association, used with permission).

of the shoreline to change over time, resulting in lateral movement of all of the previously mentioned depositional and erosional processes that occur at the shoreline. Accumulation of organic debris, including buried logs and decayed plant matter, also contributes to heterogeneity as it is incorporated with the inorganic sediments, particularly on the downwind shores of surface-water bodies. In surface-water bodies that are ice covered during winter, ice rafting during fall and spring, when ice is forming or when the ice cover is melting, can substantially rework sediments at the downwind shoreline.

3. *Biological processes*—Benthic invertebrates constantly rework sediments, particularly organic sediments, as they carry out their life cycles. Bioturbation and bioirrigation are important processes for organic sediments in deeper water environments, but it can be significant in some nearshore settings also. Aquatic birds disturb the sediment as they search for benthic invertebrates, and fish rework sediments as they create spawning redds. Beavers and muskrats can make large-scale disturbances by removing considerable amounts of sediments for lodges and passageways, and the construction of dams.

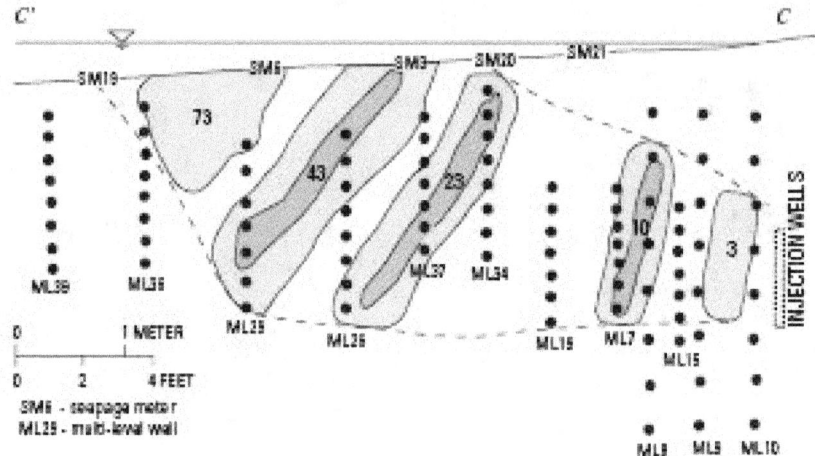

Figure 5.   Flow of salt tracer into a sandy lakebed, Perch Lake, Ontario. Numbers in shaded areas (representing center of mass of salt plume) are the days following salt injection (modified from Lee and others, 1980, copyright 1980 by the American Society of Limnology and Oceanography, Inc., used with permission).

Temporal and Spatial Variability of Flow

Flow across the sediment-water interface commonly changes in direction and velocity temporally and spatially. Many occurrences of spatial and temporal variability in the exchange between surface water and ground water are described in the literature; a few examples are provided herein. Some of this variability is summarized in figure 6. In this illustration, water flows from the surface-water body to ground water through the bottom sediments located beyond a low-permeability layer some distance from shore. Yet closer to shore, ground water flows into the surface-water body. Finally, near the shore a depression in the water table created by evapotranspiration causes flow out of the lake. At the south shoreline of Mirror Lake, New Hampshire, water flows from the lake to ground water between the shoreline and approximately 8 meters from shore, and beyond that point, flow from ground water to the lake occurs (fig. 7) (Asbury, 1990; Rosenberry, 2005).

In many settings, evapotranspiration during the summer months can depress the water table adjacent to the shoreline of wetlands, streams, and lakes below the level of the surface-water body (fig. 8) (Meyboom, 1966, 1967; Doss, 1993; Winter and Rosenberry, 1995; Rosenberry and others, 1999; Fraser and others, 2001). As a result, seasonal, and sometimes diurnal, reversals in flow between surface water and ground water may occur at the shoreline. The changes in direction of flow between surface water and ground water result from fluctuations in the amount of water removed from the water table because of evapotranspiration by plants along the margins of the surface-water body. On a seasonal basis, once evapotranspiration ceases to remove water from the near-shore regions, the near-shore depression in the water table dissipates, which then allows ground water to flow into the surface-water body.

On a diurnal basis, more evapotranspiration in the day and less at night can cause the water table to fluctuate between levels below and above the adjacent surface-water level.

In many locations, water-table mounds can develop at the edge of surface-water bodies. Many studies have shown transient water-table mounds that form in response to precipitation or snowmelt (fig. 9) (see, for example, Winter, 1986; Rosenberry and Winter, 1997; Lee and Swancar, 1997). Most of these water-table mounds were of short duration and formed in response to large rainfall events. Reversals of flow of longer duration also occur at some settings. Jaquet (1976) reported a reversal of flow along part of the shoreline at Snake Lake, Wisconsin, following spring thaw and considerable rainfall (19 centimeters) over a 5-week period that persisted for several months.

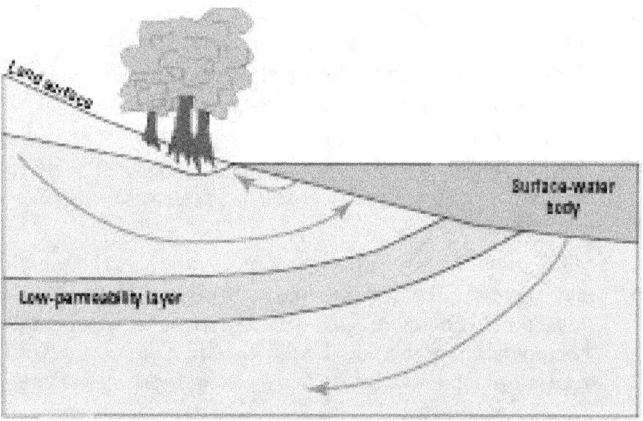

Figure 6.   The length of the flow path and the direction of flow can vary seasonally and with distance from shore. (Artwork by Donald O. Rosenberry, U.S. Geological Survey.)

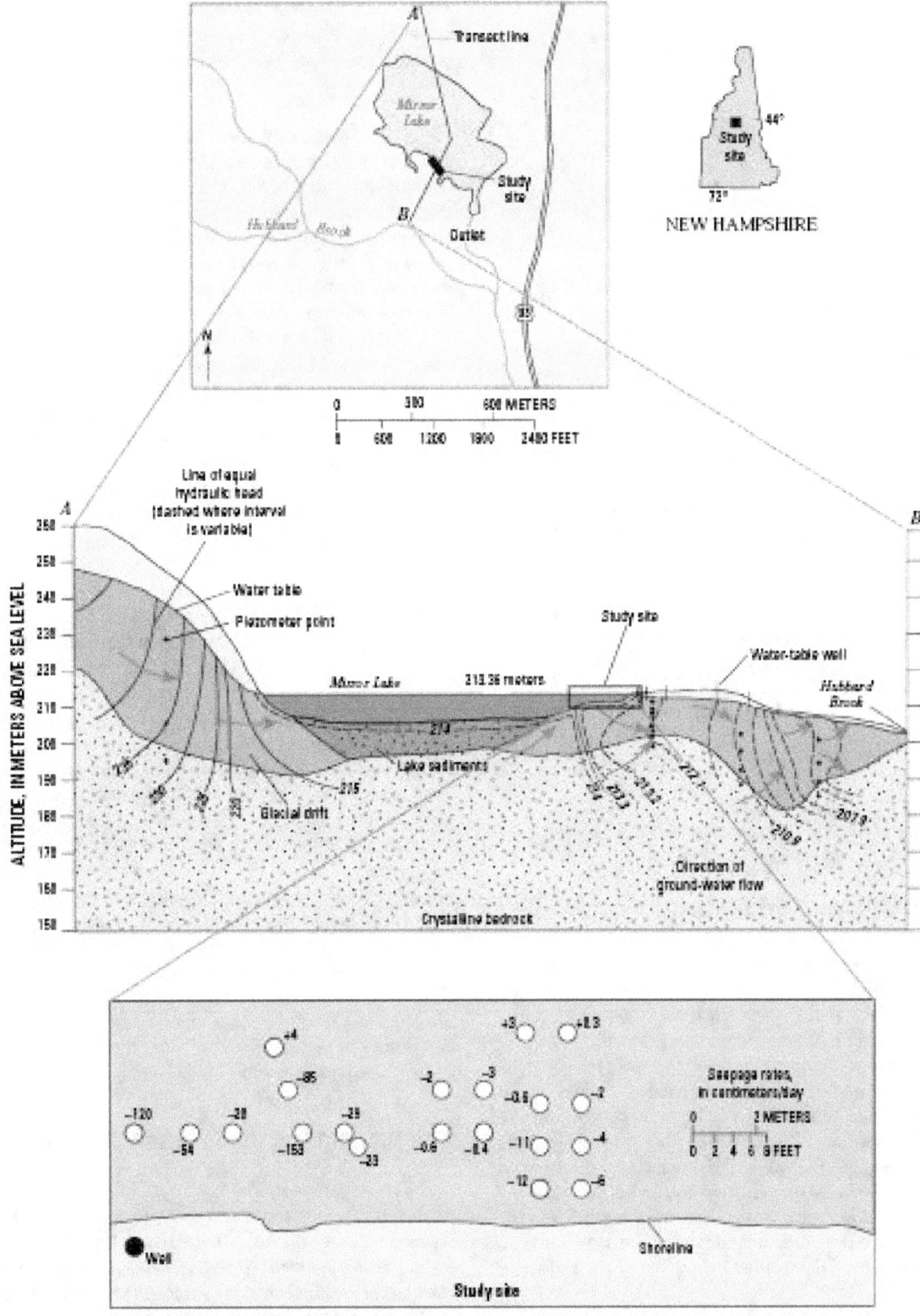

Figure 7. Example of ground-water exchange with Mirror Lake wherein lake water flows into ground water near shore (shown by negative numbers), and ground water flows into the lake farther from the shoreline (shown by positive numbers). (Modified from Rosenberry, 2005, copyright 2005 by the American Society of Limnology and Oceanography, Inc., used with permission.)

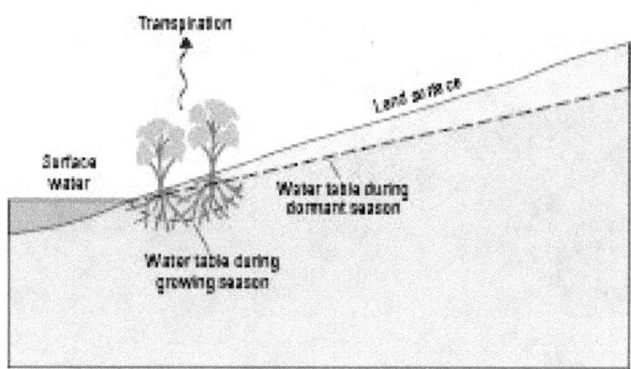

Figure 8.   Example of the effect of transpiration on the water table and the direction of water flux between surface water and ground water (from Winter and others, 1998).

Ground-water levels adjacent to streams also fluctuate in response to the rise and fall of water in the stream (Winter, 1999). An example of the resulting changes in flow direction between a stream and ground water is illustrated by data from the Cedar River in Iowa (fig. 10). In flowing waters, movement of surface water into the subsurface and out again occurs both at the bottom of the stream channel and beneath upland areas between bends in the open channel (fig. 11). This transient flow of surface water into and out of the subsurface is also known as hyporheic flow (Orghidan, 1959). Ground water flowing toward a surface-water body may discharge directly into that body or mix with hyporheic flow prior to emerging into open-water flow. The various interactions with ground water include situations in which flow is parallel to the stream (fig. 12) and does not intersect the surface water (Woessner, 1998, 2000).

Defining the Purpose for Measuring the Exchange of Water Between Surface Water and Ground Water

Water-resource investigators and water-resource managers have many reasons to quantify the flow between surface water and ground water. Perhaps the most common reasons include: calculating hydrological and chemical budgets of surface-water bodies, collecting calibration data for watershed or ground-water models, locating contaminant plumes, locating areas of surface-water discharge to ground water, improving their understanding of processes at the interface between surface water and ground water, and determining the relation of water exchange between surface water and ground water to aquatic habitat. For many investigations, it is sufficient to make a qualitative determination regarding the direction and relative magnitude of flow, either into or out of the surface-water body.

Methods for quantifying flows should be selected to be appropriate for the scale of the study. For a watershed-scale study in which multiple basins may be involved, small-scale flow phenomena, such as near-shore depressions in the water table or spatial variability of flux related to geologic

variability, likely are of little importance to the overall study goal. In such watershed-scale studies, the net flux integrated over an entire stream reach, or lake, or wetland often is the desired result. Watershed-scale flow modeling, ground-water flow modeling, flow-net analysis, or dye- and geochemical-tracer tests, often are used in such large-scale studies, studies on the order of hundreds of meters or a kilometer or more in length or breadth.

If the goal of a study is to identify and (or) delineate zones or areas of flow of surface water to ground water, or flow of ground water to surface water, smaller scale spatial and temporal variations in flow become important, and measurement tools that provide results over an intermediate scale, many tens to hundreds of meters should be selected. In many instances, measurement of surface-water flow at two places some distance apart in a segment of stream, which enables calculation of gains or losses in flow in the segment, is appropriate for these types of studies. For local, small-scale studies in which flow to or from surface water may be focused, small-scale tools such as seepage meters, small portable wells ("minipiezometers" or hydraulic potentiomanometers), and buried temperature probes may be most appropriate. Devices designed to measure flow in a small area are known as seepage meters because the term seep refers to "a small area where water moves slowly to the land surface" (USGS Water Basics Glossary http://capp.water.usgs.gov/GIP/h2o_gloss/). Seepage is defined as "the slow movement of water through small cracks, pores, interstices, and so forth, of a material into or out of a body of surface or subsurface water" (USGS Water Science Glossary of Terms http://ga.water.usgs.gov/edu/dictionary.html#S).

Once the water-resource investigator has decided on the purpose of the study and the scale of the investigation, methods of investigation can be chosen to most effectively determine where an exchange between surface water and ground water is taking place, the direction of flow, the rate or quantity of that flow, and whether the rate and direction of flow changes over time.

Determining Locations of Water Exchange

The investigator who wishes to determine where water exchange is taking place between surface water and ground water has many options, particularly in the case of ground-water discharge to surface water. Reconnaissance tools useful over larger areas, such as dye-tracer tests, aerial photography and imagery, temperature and specific-conductance probes, and surface-water discharge measurements, can be supplemented by reconnaissance tools useful in smaller areas of interest, such as seepage meters, minipiezometers, and biological indicators.

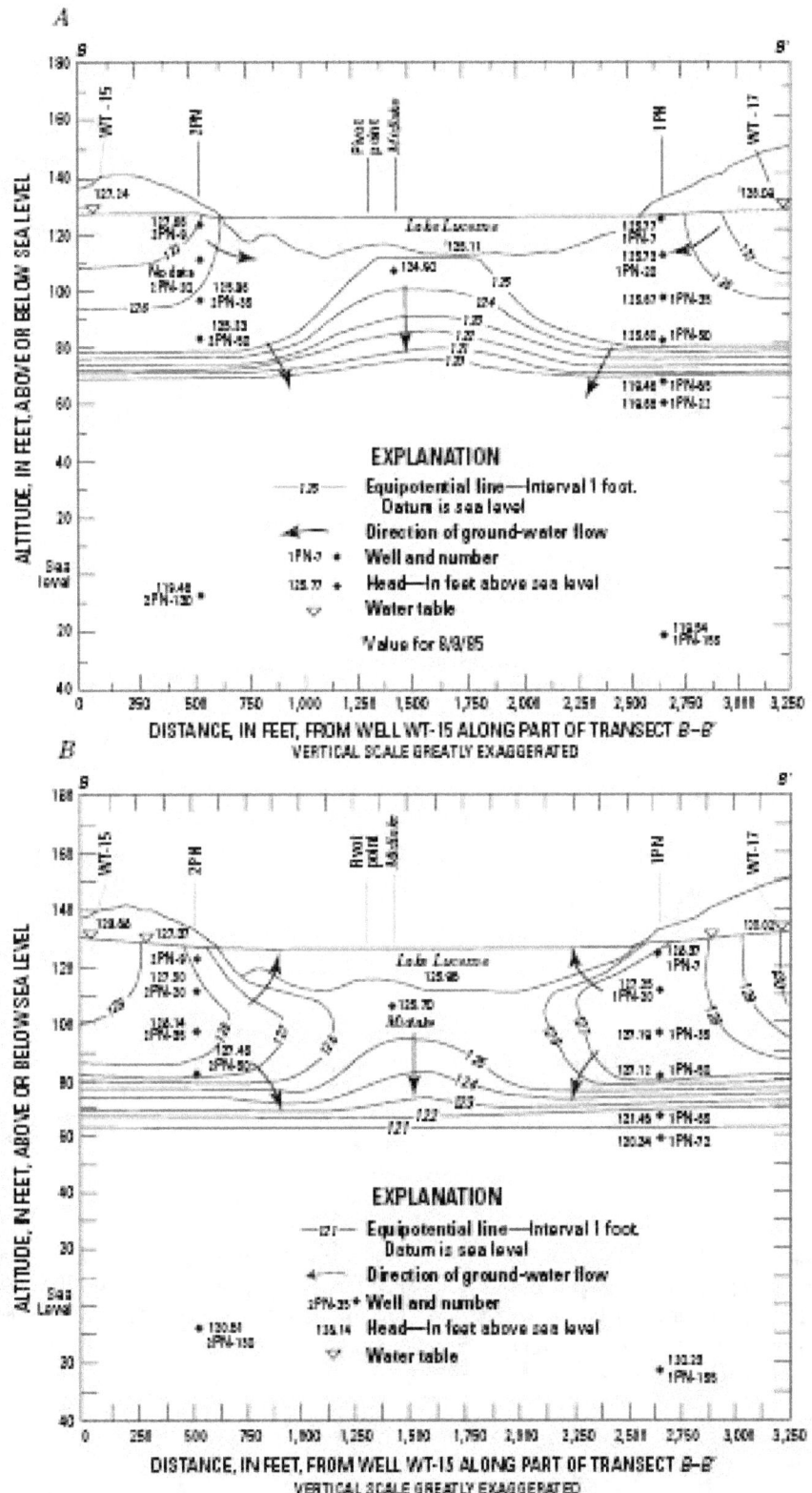

Figure 9. Example of rise and fall of water-table mounds at the edge of surface-water bodies and changes in flow direction (from Lee and Swancar, 1997). A, Vertical distribution of head showing downward head gradient conditions, August 6, 1985. B, Vertical distribution of head during high water-level conditions, October 17, 1985.

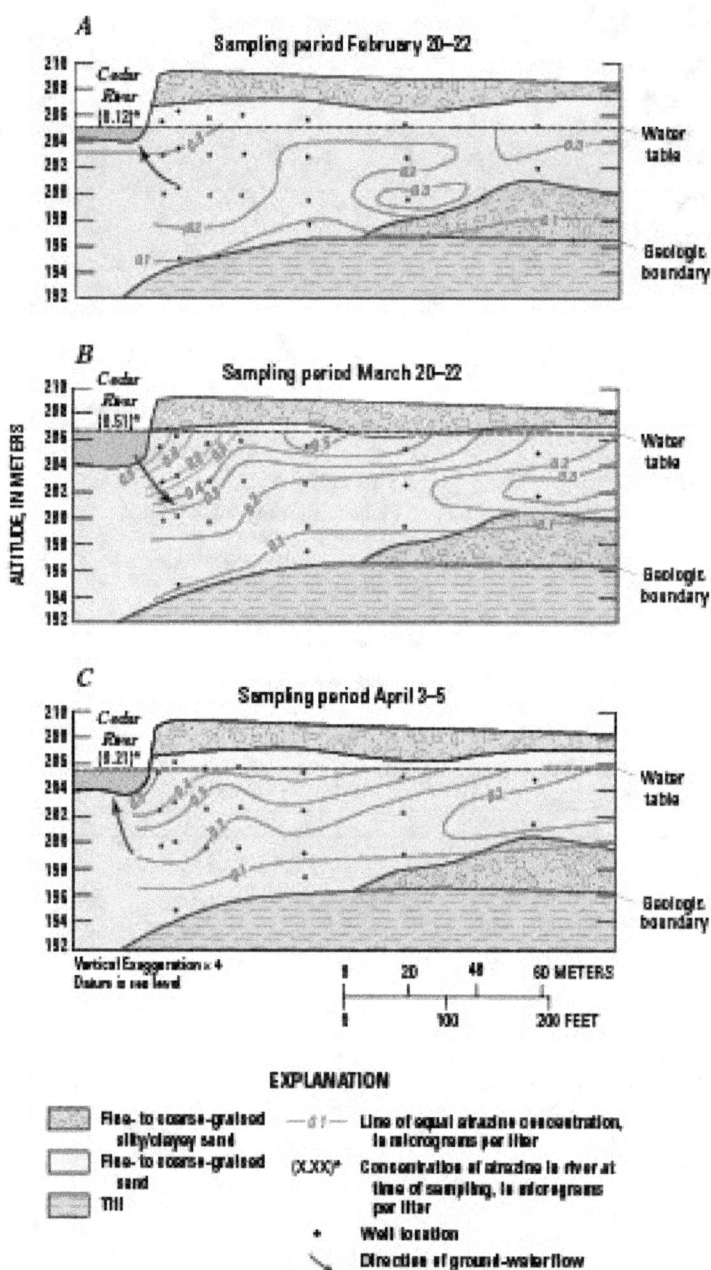

Figure 10. Example of changes in flow direction related to onset and dissipation of a water-table mound adjacent to a river (from Squillace and others, 1993, used in accordance with usage permissions of the American Geophysical Union wherein all authors are U.S. Government employees). Hydrogeologic sections for part of the Cedar River, Iowa, for three periods in 1990. *A*, Movement of ground water into the river prior to a period of high river stage. *B*, Movement of river water into the contiguous aquifer during high river stage. *C*, Return of some of the water from the aquifer during declining river stage.

A. View from Above

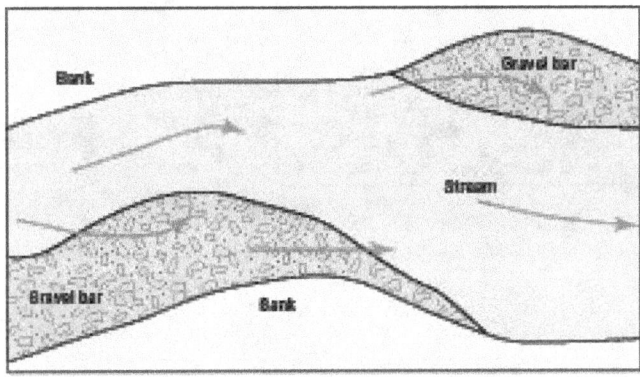

B. Sectional View

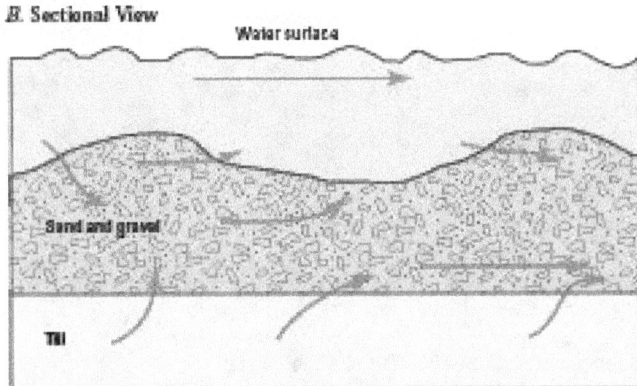

Figure 11. Example of flow interaction between surface water and ground water (from Dumouchelle, 2001). Schematic of flow in Chapman Creek, west-central Ohio. (Arrows indicate direction of flow. Diagrams not to scale.)

commonly known as a "seepage run" (Harvey and Wagner, 2000). If the amount of flow in the stream has increased over the reach, the increase may be attributed to ground-water discharge to the stream. If flow in the selected reach of stream has decreased, the decrease may be attributed to surface water flowing into ground water. It is important to recognize, however, that the direction of flow indicated by any change in streamflow is a "net direction" over the selected reach, and that within the reach, water may be moving into and out of the stream (and conversely, into and out of the underlying aquifer). It is important to account for any inflows or outflows within the stream reach, such as diversions for irrigation or channelized return flows from fields.

Measuring Direction of Flow

Comparison of surface-water levels and adjacent ground-water levels indicates direction of flow. If the surface-water level is higher than adjacent ground-water levels, the direction of flow is from the surface water to ground water. If the opposite is the case—ground-water levels are higher than nearby surface-water levels—then the direction of flow is from ground water to surface water. In addition to indicating the direction of flow, water-level measurements provide information about the magnitude of the hydraulic gradients between surface water and ground water. In some instances, however, these gradients can be altered locally. For example, vegetation between the wells and the edge of the surface-water body can transpire sufficient water to cause a local depression in the water table close to the edge of the surface water (Meyboom, 1966, 1967; Doss, 1993; Rosenberry and Winter, 1997; Fraser and others, 2001). Thus, it can be important to measure the direction of flow at a local scale using portable wells, minipiezometers, or hydraulic potentiomanometers.

Another way to determine if a section of stream or river is receiving ground-water discharge or is losing water to the underlying aquifer is by measurement of surface-water flow at two places some distance apart in a reach of stream, a practice

Measuring the Quantity of Flow

The volume of water flowing between surface water and ground water, either as surface water into ground water or ground water into surface water, can be measured directly with seepage meters. Measurement of changes in water temperatures over time at a specific site above the sediment, at the sediment-water interface, and within the sediment makes possible the determination of the amount of water exchange occurring between surface water and ground water. The exchange of water between surface water and ground water also can be examined and estimated by using dye tracer tests or by using other tracers. Such dyes or tracers are added directly to a stream and then their concentrations are measured at some point or points downstream. Changes in the concentration of the dye or other tracer over time downstream from where they are injected enables calculation of ground-water inputs.

Measuring Temporal Variations in Flow

In many instances, the rate of exchange between surface water and ground water varies over time scales of hours, days, or months. The direction of flow also may reverse on a seasonal basis or temporarily during a flood, for example. Measuring temporal variation in the rate of water exchange requires multiple measurements over these time periods. Measuring devices equipped with data recorders ("data loggers") enable the investigator to record repeated measurements at specified time intervals to document temporal changes.

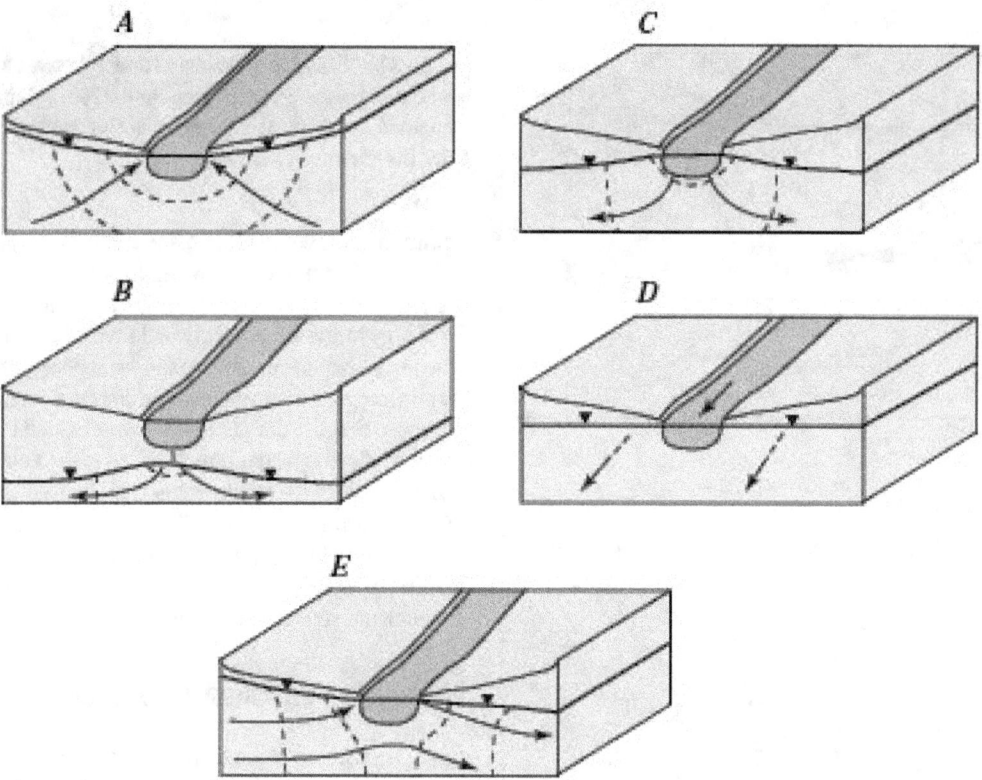

Figure 12.   Fluvial-plain ground-water and stream-channel interactions showing channel cross sections classified as: *A*, gaining; *B* and *C*, losing; *D*, zero exchange; and *E*, flow-through. The stream is dark blue. The water table and stream stage (thicker lines), ground-water flow (arrows), and equipotential lines (dashed) are shown (from Woessner, 1998, copyright American Institute of Hydrology, used with permission).

Methods of Investigation

Common methods to examine exchange of water between surface-water bodies and ground-water bodies are described below. Some of these methods make use of already installed hydrological instruments and existing data, rather than requiring the investigator to make measurements of hydrologic characteristics. When using such methods, however, the investigator may install wells, stream-gaging equipment, or rain gages, as needed, to obtain sufficient data to make the application of methods possible and the results less uncertain. Other methods require that the investigator make additional, specific measurements or observations of hydrological, physical, chemical, or biological characteristics.

Watershed-Scale Rainfall-Runoff Models

Many analytical and numerical models that relate precipitation, ground-water recharge, and ground-water discharge to temporal variability of flow in a stream have been developed. A fundamental assumption in these models is that streamflow is an integrated response to these processes over the stream's watershed, and that ground-water discharge to the stream provides the steady flow in the stream between rainfall events, commonly referred to as baseflow. Analytical models generally determine baseflow through hydrograph separation techniques. Several automated routines have been developed to assist in this determination (Rutledge, 1992; Rutledge, 1998) (fig. 13). Other analytical methods also have been used to quantify the interaction between ground water and surface water, including an analytic-element method (Mitchell-Bruker and Haitjema, 1996) and a nonparametric regression model (Adamowski and Feluch, 1991).

Several numerical models commonly referred to as rainfall-runoff models have been developed; these models areally divide watersheds and subwatersheds and calculate hydrologic parameters for each smaller area (for example, Federer and Lash, 1978; Leavesley and others, 1983, 1996, 2002; Beven and others, 1984; Beven, 1997; Buchtele and others, 1998). Rainfall-runoff models generally are calibrated to match river flow at the outlet of a watershed or subwatershed. Some models include the ground-water component of flow in each area. The current trend is to couple distributed-area watershed-scale models with ground-water flow models in order to better determine the temporal and spatial variability of the interaction between ground water and surface water (for example, Leavesley and Hay, 1998; Beven and Feyen, 2002).

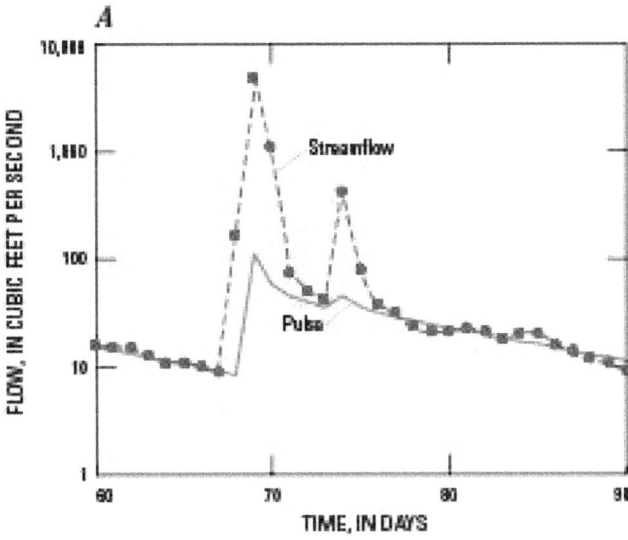

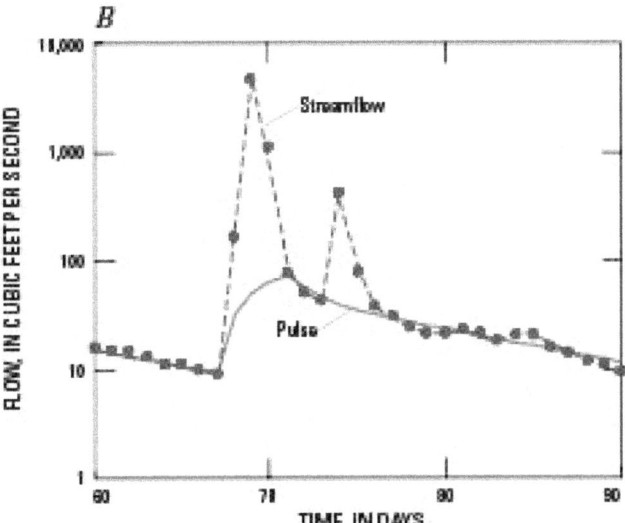

Stream Discharge Measurements

Measurements of stream discharge (Rantz and others, 1982a, b; Oberg and others, 2005) made as part of seepage runs (described earlier) can be used to determine the occurrence and rate of exchange of water between surface water and ground water in streams and rivers (fig. 14). The results of seepage runs have been used to provide an integrated value for flow between a stream and ground water along a specific stream reach. This method works well in small streams, but for larger streams and rivers, the errors associated with the measurement of flow in the channel often are greater than the net exchange of water to or from the stream or river. This method also requires that any tributaries that discharge to a stream along the reach of interest be measured and subtracted from the downstream discharge measurement. Likewise, withdrawals from the stream, such as that for irrigation, must be measured and added to the downstream discharge measurement.

Figure 13. Example of the use of hydrographs to determine the amount of ground-water discharge to a stream (from Rutledge, 2000). Hydrographs of streamflow for Big Hill Creek near Cherryvale, Kansas, for March 1974 (blue circles and dashed line), and hydrograph of estimated ground-water discharge using the PULSE model (red line). (Note: In each example, the total recharge modeled is 0.73 inch, which is the same as the total recharge estimated from RORA [a recession-curve-displacement method for estimating recharge] for this period. In example A, recharge is modeled as 0.65 inch on day 69 and 0.08 inch on day 74. In example B, recharge is modeled as a gradual process that is constant from day 68 to day 72).

The application of seepage-run data, however, is limited by the ratio of the net flow of water to or from the stream along a stream reach to the flow of water in the stream. The net exchange of water across the streambed must be greater than the cumulative errors in streamflow measurements. For example, if the errors in the stream discharge measurements are 5 percent of the true, actual flow, then according to the rules of error propagation, in order to be able to detect the net flow of water to or from the stream along the reach of interest, the value of net flow must be greater than 7 percent of the streamflow. Despite these limitations, many hydrologic studies have made use of this method with good results [for example, Ramapo River, New Jersey–Hill and others (1992); Bear River, Idaho and Utah–Herbert and Thomas (1992); Souhegan River, New Hampshire–Harte and others (1997); Lemhi River, Idaho–Donato (1998); constructed stream channel Baden-Württemberg, Germany–Kaleris (1998)]. The information gained from seepage runs can be enhanced with data obtained by using other techniques such as minipiezometers, seepage meters, temperature and specific-conductance measurements to better define surface-water/ground-water fluxes [for example, creeks and rivers in the Puget Sound area of Washington–Simonds and others (2004); and Chapman Creek, Ohio–Dumouchelle (2001)]. Seepage-run results also can provide estimates of hydraulic conductivity of the streambed on a scale appropriate for ground-water flow modeling (Hill and others, 1992).

Ground-Water Flow Modeling

Since 1983, most investigators who have used the numerical modeling approach in the quantification of flows between surface water and ground water have used the U.S. Geological Survey MODFLOW modular modeling code (Harbaugh and others, 2000). This finite-difference model contains an original "river package" that can simulate flows to or from a river, assuming the river stage does not change during a specified time period (referred to as a stress period in MODFLOW), but can change from one time period to the next. Several other MODFLOW modules or packages also have been developed to simulate fluxes between surface water and ground water. These include streamflow routing packages (Prudic, 1989;

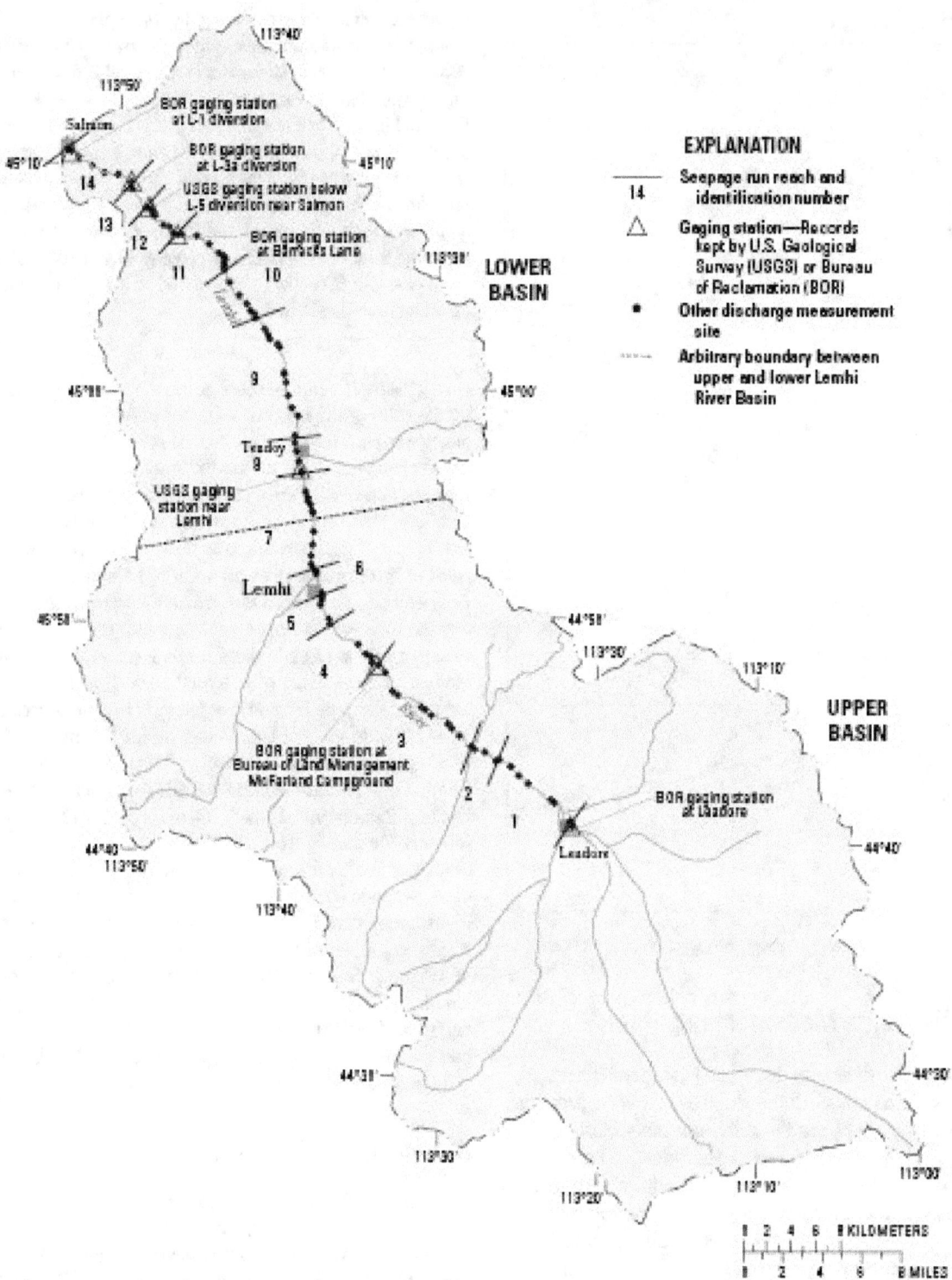

Figure 14.   Example of the use of discharge measurements to determine surface-water/ground-water interaction (from Donato, 1998). Seepage run reaches, gaging stations, and discharge measurement sites in the Lemhi River Basin, east-central Idaho, August and October 1997.

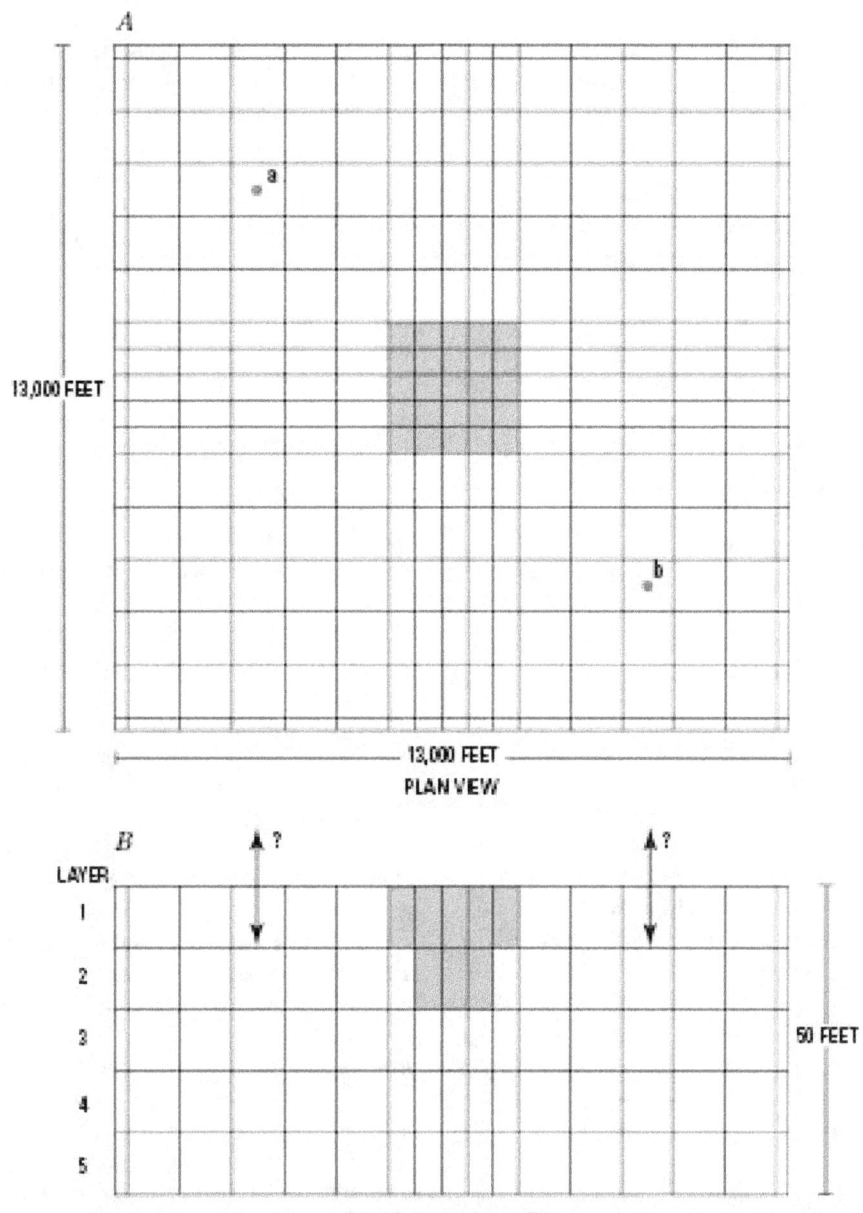

Figure 15. Example of model grid used for simulation of surface-water/ground-water interaction (from Merritt and Konikow, 2000). The lateral and vertical grid discretization for test simulation 1: *A*, Plan view—Shaded area is the surface extent of lake cells in layer 1. Interior grid dimensions are 500 and 1,000 meters. Border row/column cells are 250 feet thick. The locations of hypothetical observations wells are denoted by a and b. *B*, Cross-sectional view—Shaded area is the cross section of the lake. Although a nominal 10-foot thickness is shown for layer 1, the upper surface of layer 1 is not actually specified, and lake stages and aquifer water-table altitudes may rise higher than the nominal surface shown above.

Prudic and others, 2004), a reservoir package (Fenske and others, 1996), a lake package (Cheng and Anderson, 1993), and more recently, a more elaborate lake package (Merritt and Konikow, 2000) (fig. 15). More advanced MODFLOW-based programs have been developed to couple one-dimensional, unsteady streamflow routing with MODFLOW (Jobson and Harbaugh, 1999; Swain and Wexler, 1996). Advances also are being made in coupling MODFLOW with watershed models that simulate many of the surface-water processes within a basin (Sophocleous and others, 1999; Sophocleous and Perkins, 2000; Niswonger and others, 2006). One of the most challenging aspects of coupling ground-water and surface-water models has been representation of flow through the unsaturated zone beneath a stream. Two new programs have recently been developed for MODFLOW to simulate one-dimensional (Niswonger and Prudic, 2005) and three-dimensional (Thoms and others, 2006) flow in the unsaturated zone. Many of these packages require a determination of the transmissivity of the sediments at the interface between the aquifer and the surface-water body. Transmissivity is determined by multiplying hydraulic conductivity by the thickness of the lakebed or riverbed sediments.

Direct Measurement of Hydraulic Properties

The relation between the stage of a surface-water body and the hydraulic head measured in one or more nearby water-table wells can be used to calculate flows of water between surface water and ground water [Williams Lake, Minnesota–LaBaugh and others (1995); Vandercook Lake, Wisconsin–Wentz and others (1995); large saline lakes in central Asia–Zekster (1996); Lake Lucerne, Florida–Lee and Swancar (1997); Waquoit Bay, Cape Cod, Massachusetts–Cambareri and Eichner (1998); Otter Tail River, Minnesota–Puckett and others (2002)]. The Darcy equation (eq. 1) is used to calculate flow between ground water and surface water along specific segments of shoreline.

$$Q = KA\frac{h_1 - h_2}{L}. \tag{1}$$

where

Q is flow through a vertical plane that extends beneath the shoreline of a surface-water body (L^3/T),

A is the area of the plane through which all water must pass to either originate from the surface-water body or end up in the surface-water body, depending on the direction of flow (L^2),

K is horizontal hydraulic conductivity (L/T),

h_1 is hydraulic head at the upgradient well (L),

h_2 is hydraulic head at the shoreline of the surface-water body (L),

and

L is distance from the well to the shoreline (L).

Shoreline segments are delineated/selected on the assumption that the gradient between a nearby well and the surface-water body, the hydraulic conductivity of the sediments, and the cross-sectional area through which water flows to enter or leave the lake, are uniform along the entire segment (fig. 16). Flows through each segment are summed for the entire surface-water body to compute net flow. The scale of the shoreline segments, and the scale of the study, depend on the scale of the physical setting of interest and the density of monitoring wells. Further detail regarding this method is provided in Chapter 2, in the section "Wells and Flow-Net Analysis."

Examination and Analysis of Aerial Infrared Photography and Imagery

Aerial infrared photography and imagery have been used to locate areas of ground-water discharge to surface waters (Robinove, 1965; Fischer and others, 1966; Robinove and Anderson, 1969; Taylor and Stingelin, 1969). This technique is effective only if the temperatures of surface water and ground water are appreciably different. Information obtained from infrared scanners can be captured electronically or transferred to film, on which tonal differences correspond to differences in temperature (Robinove and Anderson, 1969; Banks and others, 1996) (fig. 17). Published studies indicate tonal differences corresponding to a difference in temperature of approximately 2 degrees Celsius are distinguishable (Pluhowski, 1972; Rundquist and others, 1985; Banks and others, 1996).

Within the limits of the ability of infrared imagery to distinguish temperature differences between surface water and ground water, the inspection of such imagery enables more rapid identification of gaining reaches in streams over large areas than can be accomplished by stream surveys that measure temperature directly (Pluhowski, 1972). Another advantage of this method in identifying areas of ground-water discharge to surface-water bodies is its application where using other techniques such as dye tracing or direct temperature measurements are impractical, or access on the ground is difficult (Campbell and Keith, 2001) or dangerous (Banks and others, 1996).

Using thermal-infrared imagery to distinguish zones of ground-water discharge is practical for locating diffuse and focused ground-water discharge (Banks and others, 1996). This capability has been demonstrated in a variety of environments. Examples for lakes are Crescent Lake, Nebraska (Rundquist and others, 1985), where the flow is diffuse and occurs over a large area, and Great Salt Lake and Utah Lake in Utah (Baskin, 1998), where the ground-water flow into the lake is focused at springs. Campbell and Keith (2001) found the technique useful in locating many springs flowing into streams and reservoirs in northern Alabama. Examples for estuaries are creeks flowing into Chesapeake Bay, Maryland, and the shorelines of the Gunpowder River and the Chesapeake Bay into which the river flows (Banks and others,

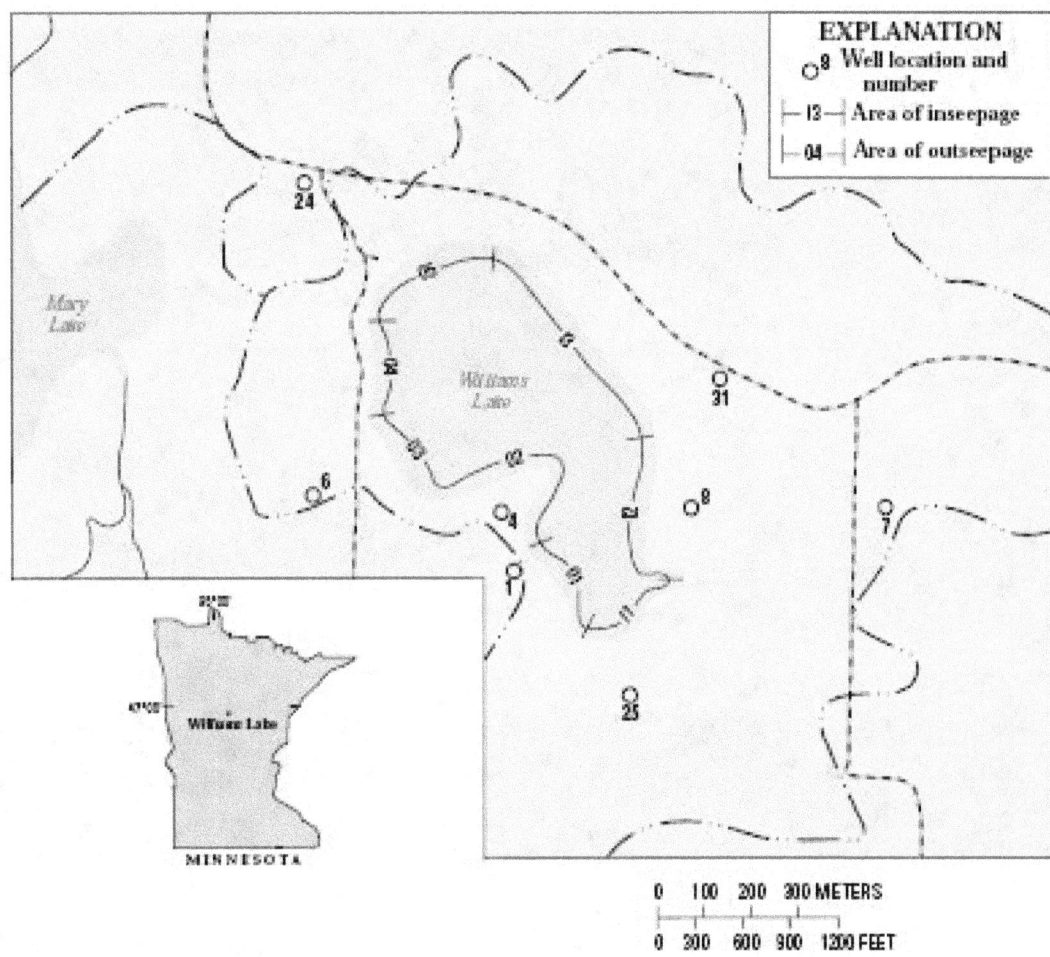

Figure 16.   Example of shoreline segment definition for the calculation of water fluxes between surface water and ground water at a lake (modified from LaBaugh and others, 1995, used in accordance with author rights of the National Research Council of Canada Press). Location of wells and shoreline segments used to calculate flow between surface water and ground water at Williams Lake, Minnesota.

1996), as well as creeks and rivers flowing into Long Island Sound, New York (Pluhowski, 1972). Examples of the use of infrared imagery to detect areas of ground-water discharge to marine waters include the delineation of areas of diffuse ground-water flow into Long Island Sound (Pluhowski, 1972) and focused ground-water flow as springs to the ocean, such as around the perimeter of the island of Hawaii (Fischer and others, 1966).

Dye and Tracer Tests

Dyes and other soluble tracers can be added to water and then "tracked" to provide direct, qualitative information about ground-water movement to streams. Fluorescent dyes that are readily detected at small concentrations and pose little environmental risk make a useful tool for tracing ground-water flow paths, particularly in karst terrane (Aley and Fletcher, 1976; Smart and Laidlaw, 1977; Jones, 1984; Mull and others,

1988). Thus, dye-tracer studies can be used to determine the time-of-travel for ground water to move to and into surface water, as well as hydraulic properties of aquifer systems (Mull and others, 1988). The use of dyes as tracers is described in more detail in Chapter 3. Commonly, a reconnaissance of the ground-water basin is made to identify likely areas of potential surface-water flow into ground water or ground-water flow to the surface. An inventory is made of springs, sinkholes, boreholes or screened wells, and sinking streams. Appropriate sites then are picked for dye injection, and the potential discharge areas, springs, and stream reaches are monitored over an appropriate period of time, hours or days, for appearance of the dye (fig. 18).

Solute tracers have been used to aid in the determination of water gains or losses within the channel of a stream or river (Kilpatrick and Cobb, 1985). This technique is known as dilution-gaging. A variety of tracers have been used in such studies, either alone or in combination, usually including a

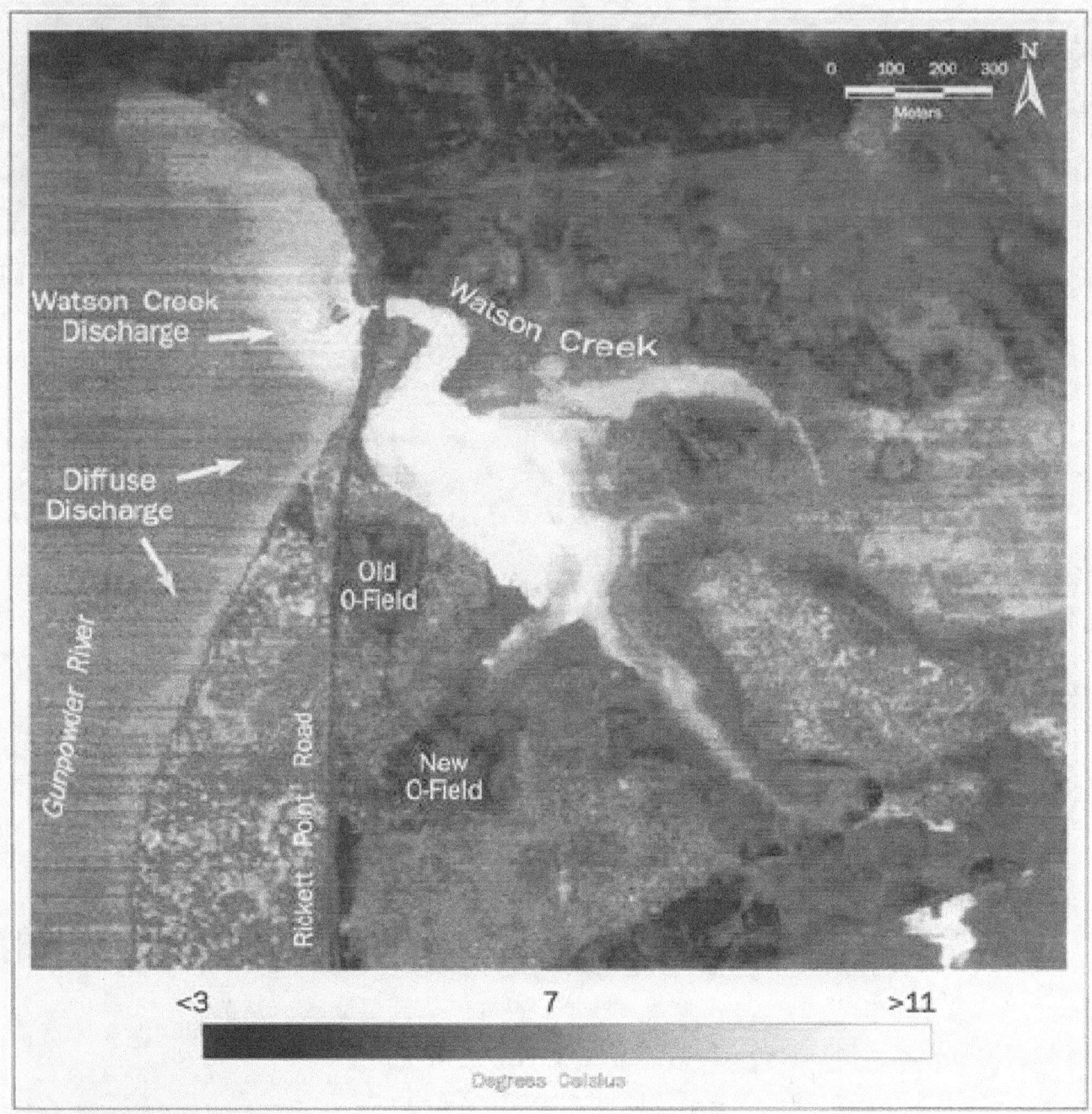

Figure 17.   Example of use of thermal infrared imagery to delineate areas of discrete and diffuse ground-water discharge to surface water (from Banks and others, 1996, reprinted from Ground Water with permission from the National Ground Water Association, copyright 1996, thermal imagery of O-Field study area, Aberdeen Proving Ground, Maryland).

Figure 18.   Example of the use of dye to examine water fluxes between surface water and ground water (from Carter and others, 2002). Dye testing has been done in Boxelder Creek, South Dakota, which can lose as much as 50 cubic feet per second of flow to the bedrock aquifers. In the upper left photograph, nontoxic, red dye is poured into Boxelder Creek upstream from a major loss zone. In the upper right photograph, dye in the stream can be seen disappearing into a sinkhole in the Madison Limestone. In the bottom photograph, dye in the stream emerges downstream at Gravel Spring, which is about 671 meters (2,200 feet) (linear distance) from the major loss zone. The length of time for the first arrival of dye to travel this distance is variable depending on flow conditions but generally is about 1 to 2 hours (Strobel and others, 2000). Thus, the ground-water velocity is about 0.3 to 0.6 kilometer per hour (0.2 to 0.4 mile per hour), which is a very fast rate for ground water. Dye also has been recovered at City Springs, which is in the Rapid Creek Basin, about 30 days after injection. This demonstrates that ground-water flow paths are not necessarily restricted by surface-water drainage basins. (Photographs by Derric L. Isles, South Dakota Department of Environment and Natural Resources.)

tracer expected to be nonreactive in the waters of the stream to which it is added, such as lithium (for example, in the Snake River, Colorado–Bencala and others, 1990) or chloride (for example, Chalk Creek, Colorado–Kimball, 1997). In this technique, a known quantity of solute is added at a specified rate for a short interval of time at the upstream cross section of the stream segment of interest, and concentrations of the solute are measured at one or more points downstream over time. Discharge is calculated from the amount of dilution that occurs at the downstream point or points. Kimball (1997) indicated that Q_s (the discharge in the stream) is calculated as follows:

$$Q_s = (C_i Q_i) / (C_B - C_A) \qquad (2)$$

where

C_i is the tracer concentration in the injection solution,

Q_i is the rate of injection into the stream,

C_B is the tracer concentration downstream,

and

C_A is the tracer concentration upstream from the injection point.

When coupled with stream-segment discharge measurements, use of solute tracers also enables calculation of the rates of ground-water inflow and outflow within a stream segment (Harvey and Wagner, 2000). At the same time the solute is injected and monitored within the stream segment, physical velocity measurements (streamflow) are made at the upstream and downstream sections of the stream reach. The streamflow measurements provide information on whether or not there was a net loss or gain of flow within the reach due to interaction with ground water. Harvey and Wagner (2000) indicate the solute tracer, or dilution-gaging, values determine ground-water inflow. Thus, ground-water outflow can be calculated by subtracting the net loss or gain from the solute tracer-derived ground-water inflow value.

Calculation of chemical budgets for a stream, lake, or wetland is another way in which solutes can be used to make quantitative estimates of surface-water exchange with ground water. Conservative chemicals in a watershed are those that are not altered by the porous media through which they flow, and occur at concentrations for which changes in concentration because of chemical precipitation are not likely to occur. Conservative chemicals can be used to determine the volume of ground water that flows into or out of a surface-water body, provided that all other fluxes are known. A common form of the chemical-budgeting equation for a lake or wetland is

$$P(C_P) + GWI(C_{GWI}) + SI(C_{SI}) - GWO(C_{GWO})$$
$$- SO(C_{SO}) = \Delta V_L(C_L) \pm R. \qquad (3)$$

where

P is precipitation,

GWI is ground-water flux into lake or wetland,

SI is streamflow into lake or wetland,

E is evaporation,

GWO is flux of lake or wetland water to ground water,

SO is streamflow out of lake or wetland,

ΔV_L is change in lake or wetland volume,

C_n is chemical concentration of hydrologic component,

and

R is residual.

Chemical budgets have been calculated in lake and wetland studies where water exchange between surface-water bodies and underlying ground water was of interest [see, for example, Rawson Lake, Ontario–Schindler and others (1976); Thoreau's Bog, Massachusetts–Hemond (1983); Williams Lake, Minnesota–LaBaugh and others (1995); LaBaugh and others (1997); multiple lakes in Polk and Highlands Counties, Florida–Sacks and others (1998); Lake Kinneret, Israel–Rimmer and Gideon (2003)]. The equation can be modified to solve for any unknown flow term, provided that the remainder of the flow terms are known. The accuracy of the method depends greatly on the accuracy of the other flow and chemical-concentration measurements. The size of the residual term often is considered a general indicator of the accuracy of the method, but a small residual does not always indicate an accurate chemical balance. LaBaugh (1985) and Choi and Harvey (2000) provide examples of the use of error analysis to quantify the uncertainty associated with water-flux results obtained using this method.

The ratios of the isotopes of oxygen and hydrogen present in water have been used for decades to distinguish sources of water, including ground-water discharge to surface-water bodies (for example, Dincer, 1968). These isotopes are useful because they are part of the water and not solutes dissolved in the water. The method works well when the degree of isotopic fractionation of the water is different for different sources of water. The process of evaporation tends to remove lighter isotopes, leaving the heavier isotopes behind. Thus, the ratio of lighter to heavier isotopes will change over time in the water and the water vapor. More detailed explanation of the isotopic fractionation in catchment water is given in Kendall and others (1995). If the isotopic compositions of different sources of water are distinct, then simple mixing models can be used to identify sources of water. A brief example is presented here, but more detailed explanations and examples of the use of this method can be found in Krabbenhoft and others (1994), LaBaugh and others (1997), Sacks (2002) (applied to lakes), Kendall and others (1995) (a brief description of the methods), and Kendall and McDonnell (1998) (detailed descriptions of numerous isotopic methods).

For determination of ground-water discharge to streams and rivers, a simple two-component mixing model often is used:

$$Q_S \delta_S = Q_{GW} \delta_{GW} + Q_P \delta_P \qquad (4)$$

where

Q is discharge,

δ is the stable-isotopic composition in parts per thousand enrichment or depletion ("per mil") relative to a standard,

S is stream water,

GW is ground water,

and

P is precipitation.

For lakes and wetlands, where sources of water are more numerous, slightly more complex mixing models can be used, such as those provided by Krabbenhoft and others (1990):

$$GWI = \frac{P(\delta_L - \delta_P) + E(\delta_E - \delta_L)}{\delta_{GWI} - \delta_L} \qquad (5)$$

or that provided by Krabbenhoft and others (1994):

$$GWO = \frac{P \cdot \delta_P - \Delta V_L \cdot \delta_L - GWI \cdot \delta_{GWI}}{\delta_L}. \qquad (6)$$

where

GWI is ground-water flux into lake,

GWO is flux of water from lake to ground water,

P is precipitation,

E is evaporation,

ΔV_L is change in the volume of the lake,

and

δ_x is per mil value for hydrologic component.

Where equations 5 and 6 are derived from equation 3 applied to stable isotopes at steady state:

$$d(V\delta_L)/dt = GWI\, \delta_{GW} + P\, \delta_P + Si\, \delta_{Si}$$
$$- GWO\, \delta_L - E\, \delta_E - So\, \delta_L = 0. \qquad (7)$$

Investigation of ground-water discharge into inland and marine surface water also is feasible through measurement of radon and radium isotopes (Corbett and others, 1998; Moore, 2000). In the radon isotope method, a mass balance is constructed for radon-222 (^{222}Rn), which is a chemically and biologically inert radioactive gas formed by the disintegration of the parent nuclide radium (Corbett and others, 1998, 1999). Because radon is a gas, radon in water in contact with the atmosphere will be lost from that water because of volatilization. Thus, ground water commonly contains higher activities of ^{222}Rn than does surface water, from 3 to 4 orders of magnitude greater (Burnett and others, 2001). ^{222}Rn is a radioactive daughter isotope of radium 226 (^{226}Ra) and has a half life of 3.82 days. Determination of the activity of ^{222}Rn and ^{226}Ra in surface water enables the calculation of the ^{222}Rn excess—how much more ^{222}Rn is present in surface water than would be expected based on the ^{226}Ra content of the water. Determination of the activity of ^{222}Rn and ^{226}Ra in sediment water or ground water is used to determine the ^{222}Rn flux into surface water (Cable and others, 1996; Corbett and others, 1998), which can account for the excess ^{222}Rn in the surface water. The mass balance or flux of ^{222}Rn has been used to determine ground-water discharge to several types of surface-water bodies (Kraemer and Genereaux, 1998): in streams, such as the Bickford watershed, Massachusetts (Genereaux and Hemond, 1990); rivers, such as the Rio Grande de Manati, Puerto Rico (Ellins and others, 1990); lakes, such as Lake Kinneret, Israel (Kolodny and others, 1999); estuaries, such as Chesapeake Bay (Hussain and others, 1999), Charlotte Harbor, Florida (Miller and others, 1990), and Florida Bay (Corbett and others, 1999), as well as the coastal ocean, such as in Kanaha Bay, Oahu, Hawaii (Garrison and others, 2003); the Gulf of Mexico off of Florida (Cable and others, 1996; Burnett and others, 2001); and the Atlantic Ocean off the coast of South Carolina (Corbett and others, 1998).

In the radium isotope method, the surface-water activities of the four naturally occurring radium isotopes—^{226}Ra, ^{228}Ra, ^{223}Ra, and ^{224}Ra—are compared to activities in sediment water or ground water to determine fluxes (Moore, 2000). The source of the radium isotopes is the decay of uranium and thorium in sediments or rocks. Water in contact with solid materials containing the source of the isotopes will accumulate the isotopes. Ground water will accumulate more of the isotopes because of water's presence within the matrix of the sediments or rocks. Surface waters will accumulate less of the isotopes because the sediments or rocks are less abundant relative to the water (Kraemer, 2005). Kraemer indicates the ratio of the longer lived isotopes (^{226}Ra half-life of 1,601 years, ^{228}Ra half-life of 5.8 years) can be used as an indicator of the types of sediments or rocks through which ground water has traveled, because of differences in uranium and thorium content between rock types. Kraemer (2005) also notes that the short-lived isotopes (^{223}Ra half life of 11.4 days, and ^{224}Ra half-life of 3.7 days) provide some indication of the timing of ground-water discharge. Naturally occurring radium isotopes have been useful in the identification of ground-water inflow to lakes, such as Cayuga Lake, New York (Kraemer, 2005), freshwater wetlands in the Florida Everglades (Krest and Harvey, 2003), estuarine wetlands, North Inlet salt marsh, South Carolina (Krest and others, 2000), and coastal waters in the central South Atlantic Bight (Moore, 2000).

Thermal Profiling

Measurements of temperature have been used to determine qualitatively the locations of rapid discharge of ground water to surface water; common measurement methods include towing a tethered temperature probe, in-situ measurements of temperature, and thermal imagery (Lee, 1985; Baskin, 1998; Rosenberry and others, 2000). Temperature also can be measured at different depths beneath a stream or lake to determine the rate of vertical flow through a surface-water bed either into or out of the surface-water body (Lapham, 1989). This method is effective when flow through the lakebed is sufficient to allow advective processes to be significant relative to conductive temperature signals. The method requires multiple measurements of temperature over weeks or months and the simultaneous solution of the flow of fluid and heat in one dimension. Depths of temperature measurement typically extend up to 3 to 6 meters beneath the sediment bed. The solution requires the assumption that flow is vertical through the surface-water bed and that the media are homogeneous and isotropic. Taniguchi (1993) used seasonal changes in sediment temperature beneath a surface-water body to develop type curves that can be used to estimate vertical fluxes through the surface-water bed.

Subsediment temperature has been used over short distances beneath the surface-water bed and makes use of the temperature response in the sediments to diurnal changes in surface-water temperature (Constantz and others, 1994; Stonestrom and Constantz, 2003). This method adjusts parameters in a one-dimensional, variably saturated heat-transport model (VS2DH) [developed in the USGS (Healy and Ronan, 1996)] until the simulation results match the temperature data collected beneath the surface-water body. Sediment temperature measurements and data also have been used to determine streamflow frequency in ephemeral stream channels (Constantz and others, 2001; Stonestrom and Constantz, 2003). This method is described in greater detail in Chapter 4. Conant (2004) used measurements of hydraulic-head gradient and hydraulic conductivity in wells installed in a streambed to determine rates of exchange between ground water and surface water. Conant then developed an empirical relation between streambed flux and streambed temperature relative to stream-water temperature

and used the empirical relation to indicate rate of discharge of ground water to the stream at locations where only streambed temperature was measured.

Use of Specific-Conductance Probes

Specific-conductance probes are another tool that can be useful for locating areas of ground-water discharge to surface waters (Lee, 1985; Vanek and Lee, 1991; Harvey and others, 1997) (fig. 19). Such probes are suspended from a boat with a cable connecting the probes to a specific-conductance meter on board the boat. The housing for the electrically conductive probes is designed to maintain contact with the sediments (Lee, 1985) so that the probes are dragged through bottom sediments at a depth of 1 to 3 centimeters (Vanek and Lee, 1991). This method depends on having the existence of a difference in the electrical conductance of surface and ground water great enough to be detected by the sensors (probes) and thus to identify points or areas of ground-water discharge to the surface water. Thus, saline waters receiving ground-water discharge of less salinity (Vanek and Lee, 1991) or fresh surface waters receiving more mineralized ground-water discharge (Harvey and others, 1997) are environments where this technique is effective. Changes in electrical conductance, however, also may reflect changes in sediment type so that the technique should be considered only as a reconnaissance tool. The identification of places where ground-water discharge may be occurring should be verified by other, complementary techniques (Vanek and Lee, 1991). Specific-conductance and temperature probes have been used in this way along many kilometers of river reaches in combination with Global Positioning System information to determine the precise locations of ground-water discharge and their variation with season (Vaccaro and Maloy, 2006).

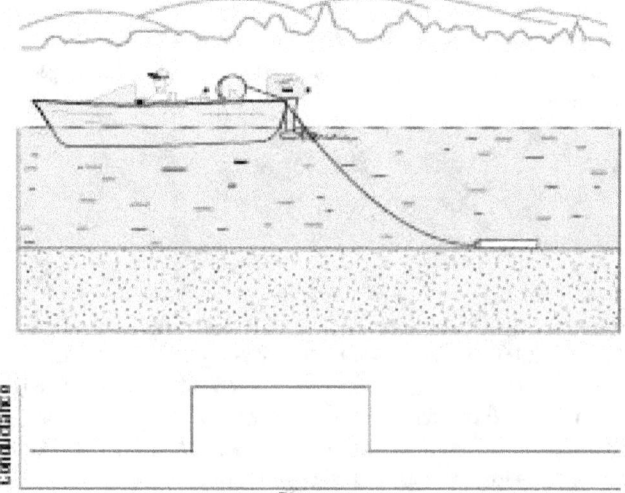

Figure 19.   Example of the use of a towed specific-conductance probe to identify ground-water discharge to surface waters (from Harvey and others, 1997, used in accordance with usage permissions of the American Geophysical Union). The sediment probe is being towed behind a small boat. The probe is used to detect areas of more electrically conductive ground-water inflow. In saline waters, the probe is used to detect less electrically conductive fresh ground-water inflow.

Electrical Resistivity Profiling

Electrical geophysical survey methods are applicable to delineating submarine ground-water discharge (Loke and Lane, 2004). Salinity differences in water can be detected by the resistance to the passage of current between electrodes. Fresh or weakly saline waters will be more resistant to movement of an electric current than more saline waters. Two resistivity data acquisition geometries commonly are used: (1) continuous resistivity profiling (Belaval and others, 2003; Manheim and others, 2004; Day-Lewis and others, 2006), and (2) marine resistivity (Taniguchi and others, 2006). In continuous resistivity profiling, a streamer comprising a set of floating electrodes spaced at a regular interval is towed on the water surface behind a boat. As the boat travels along a transect, electrical current is applied at a fixed time interval at one or more electrode pairs, and electrical potentials are measured simultaneously between other electrode pairs. At the same time, water depths along the transect are measured with echo sounding. In marine resistivity profiling, measurements are made using electrodes placed on the water bottom. Regardless

of the acquisition geometry, data are inverted to produce two-dimensional cross sections, or tomograms, of subsurface resistivity. Such data can indicate locations within a transect where submarine ground-water discharge is occurring, as well as delineating the subsurface saltwater/freshwater interface or geologic structure. Tomograms are commonly interpreted in the context of direct, discrete measurements of conductivity, temperature, and depth.

Hydraulic Potentiomanometer (Portable Wells) Measurements

Many devices have been designed to be installed through the bed of a surface-water body to measure the vertical hydraulic-head gradient beneath the surface-water body. These devices provide a direct measurement of hydraulic head relative to surface-water stage at the depth to which the probe is inserted beneath the surface-water bed. One of the most commonly used devices of this type is the hydraulic potentiomanometer (fig. 20), also sometimes called a minipiezometer

A

B

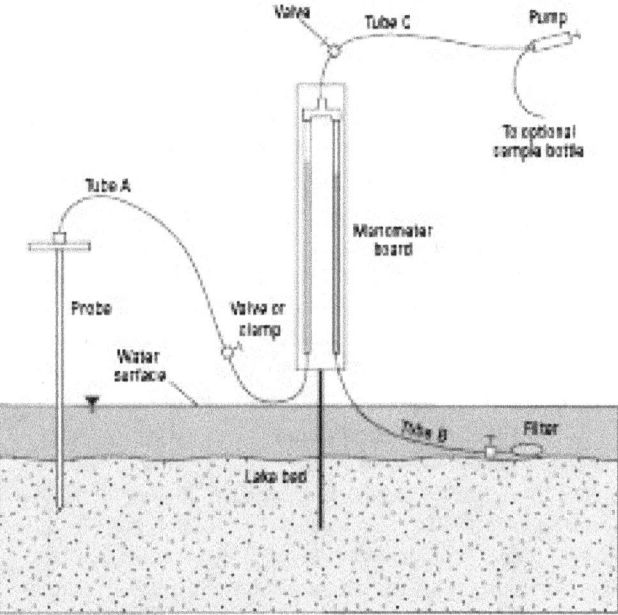

Figure 20. *A*, Hydraulic potentiomanometer with users demonstrating hydraulic-head difference between surface water and ground water. The hydraulic head of the river (right side of the board) is higher than the hydraulic head of the ground water (left side of the board), indicating a downward gradient (from Simonds and Sinclair, 2002). Photograph by Kirk A. Sinclair, Washington State Department of Ecology, 2002. *B*, Components of a hydraulic-potentiomanometer system (from Winter and others, 1988, copyright 1988 by the American Society of Limnology and Oceanography, Inc., used with permission).

(Winter and others, 1988). Although this device does not provide direct measurements of flow across the sediment-water interface, it is useful for making qualitative determinations of direction of flow across the sediment-water interface. Estimates of hydraulic conductivity also can be made on the basis of the amount of suction required to pull water through the screen at the end of the probe. These devices often are used in conjunction with a network of wells to obtain additional hydraulic-head data along shoreline segments where wells are not or cannot be situated. Details about use of these devices is provided in Chapter 2.

Seepage Meter Measurements

Seepage meters are devices that isolate a small area of the bed of a surface-water body and measure the flow of water across that area. References as early as 1944 indicate that early seepage meters were designed to measure water loss through unlined canals (Carr and Winter, 1980). Beginning in the early 1970s, seepage meters have been used in lakes, rivers, wetlands, and estuaries to measure flows between surface water and ground water in natural settings. Use in coastal environments has increased during recent years (Cable and others, 1997). One of the most common devices, called a half-barrel seepage meter, uses a cut-off end of a steel (or plastic) storage drum to isolate a small circular area of the surface-water bed, and a plastic bag is attached to the barrel to register the change in water volume over the time of bag attachment (Lee, 1977; Lee and Cherry, 1978) (fig. 21). More details about the use and interpretation of data from seepage meters are included in Chapter 2. Many device modifications have been made to the basic design of seepage meters for use in deep water, soft sediments, shallow water, and areas exposed to large waves.

Many investigations of surface-water and ground-water interaction require integrating point measurements of water flux in order to interpret the total flow between surface water and ground water. The most common method is to average point measurements and apply that average value to all or part of the surface-water body of interest. This method is not appropriate, however, where ground-water flux into a surface water declines with greater distance from shore. If data are collected along transects perpendicular to the shore, a curve can be fit through the data. Because ground-water flow to surface water commonly is distributed exponentially with distance from shore, exponential curves have been fit through transect data, and the equation for the curve has been used to calculate water flux for an area deemed to be representative of the transect point measurements (for example, Fellows and Brezonik, 1980). Where ground-water flux is not exponentially distributed with distance from shore, a plot of point measurements of seepage flux with distance from shore is made, and the area under the curve can be determined to represent total seepage for a unit width of shoreline. That value can

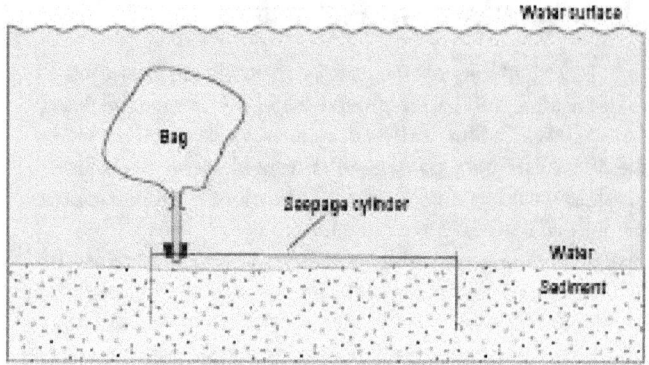

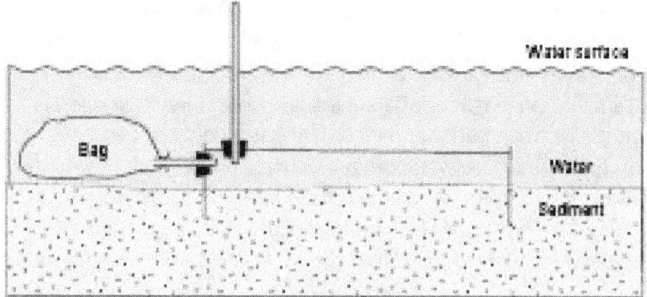

Figure 21. Full-section view of seepage meter showing details of placement in the sediment (modified from Lee and Cherry, 1978, used with permission of the Journal of Geoscience Education).

then be multiplied by an appropriate shoreline length to determine a flux volume for a specific area of the surface-water body. If point measurements are made at a sufficient number of transects, ground-water flux for an area of a surface-water body can be determined.

Biological Indicators

The biological response to conditions of flow at the sediment-water interface can be an indicator of the direction and relative magnitude of flow. The growing field of ground-water ecology has made frequent use of distributions of specific types of plants and animals as an indicator of ground-water/surface-water interaction (Danielopol, 1984; Danielopol and others, 1997; Lodge and others, 1989; Malard and others, 1996; Goslee and others, 1997; Wetzel, 1999) (fig. 22). These methods are useful reconnaissance tools to aid in locating areas in need of more detailed investigations. Typically, these methods involve identifying species or groups of species of plants or animals that are known to thrive in places where ground water discharges to surface water, but some of the indicators also indicate areas where surface water flows into ground water. Identification of specific plant and animal species is necessary for use of these methods, but some of the species are readily identifiable without requiring intensive biological or ecological training (Rosenberry and others, 2000).

Figure 22.   Marsh marigold in Shingobee Lake, Minnesota. Presence indicates location of ground-water discharge to the lake. (Photographs by Donald O. Rosenberry, U.S. Geological Survey.)

References

Adamowski, K., and Feluch, W., 1991, Application of nonparametric regression to groundwater level prediction: Canadian Journal of Civil Engineering, v. 18, no. 4, p. 600–606.

Aley, T.J., and Fletcher, M.W., 1976, The water tracer's cookbook—A guide to the art and science of water tracing materials with particular emphasis on the use of fluorescent dyes, Lycopodium spores, and bacteriophage in groundwater investigations: Missouri Speleology, v. 16, no. 3, 32 p.

Alley, W.M., Reilly, T.E., and Franke, O.L., 1999, Sustainability of ground-water resources: U.S. Geological Survey Circular 1186, 79 p.

Andrews, C.B., and Anderson, M.P., 1978, Impact of a power plant on the ground-water system of a wetland: Ground Water, v. 16, no. 2, p. 105–111.

Asbury, C.E., 1990, The role of groundwater seepage in sediment chemistry and nutrient budgets in Mirror Lake, New Hampshire: Ithaca, Cornell University, Ph.D. dissertation, 275 p.

Attanayake, M.P., and Waller, D.H., 1988, Use of seepage meters in a groundwater-lake interaction study in a fractured rock basin—A case study: Canadian Journal of Civil Engineering, v. 15, p. 984–989.

Bacchus, S.T., 2001, Knowledge of groundwater responses—A critical factor in saving Florida's threatened and endangered species, Part 1—Marine ecological disturbances: Endangered Species UPDATE, v. 18, no. 3, p. 79–90.

Bacchus, S.T., 2002, The "ostrich" component of the multiple stressor model—Undermining South Florida, in Porter, J.W., and Porter, K.G., eds., The Everglades, Florida Bay, and Coral Reefs of the Florida Keys—An Ecosystem Sourcebook: Boca Raton, CRC Press, p. 677–748.

Banks, W.S.L., Paylor, R.L., and Hughes, W.B., 1996, Using thermal-infrared imagery to delineate ground-water discharge: Ground Water, v. 34, no. 3, p. 434–443.

Barlow, P.M., and Dickerman, D.C., 2001, Numerical simulation and conjunctive-management models of the Hunt-Annaquatucket-Pettaquamscutt stream-aquifer system, Rhode Island: U.S. Geological Survey Professional Paper 1636, 88 p., one map sheet.

Barlow, P.M., and Wild, E.C., 2002, Bibliography on the occurrence and intrusion of saltwater in aquifers along the Atlantic Coast of the United States: U.S. Geological Survey Open-File Report 2002-235, 30 p.

Baskin, R.L., 1998, Locating shoreline submarine springs in two Utah lakes using thermal imagery, in Pitman, J.K., and Carroll, A.R., eds., Modern & ancient lake systems—New problems and perspectives: Salt Lake City, Utah Geological Association, v. 26, p. 51–57.

Belanger, T.V., and Kirkner, R.A., 1994, Groundwater/surface water interaction in a Florida augmentation lake: Lake and Reservoir Management, v. 8, no. 2, p. 165–174.

Belaval, Marcel, Lane, J.W., Jr., Lesmes, D.P., and Kineke, G.C., 2003, Continuous-resistivity profiling for coastal groundwater investigations—Three case studies, in Symposium on the Application of Geophysics to Engineering and Environmental Problems (SAGEEP), April 6–10, 2003, San Antonio, Texas, Proceedings: Denver, Colorado, Environmental and Engineering Society, CD-ROM, 14 p.

Bencala, K.E., McKnight, D.M., and Zellweger, G.W., 1990, Characterization of transport in an acidic and metal-rich mountain stream based on a lithium tracer injection and simulations of transient storage: Water Resources Research, v. 26, no. 5, p. 989–1000.

Beven, K.J., ed., 1997, Distributed hydrological modelling—Application of the TOPMODEL concept: Chichester, Wiley, 348 p.

Beven, K.J., and Feyen, J., 2002, The future of distributed modeling: Hydrological Processes, v. 1, p. 169–172.

Beven, K.J., Kirkby, M.J., Schofield, N., and Tagg, A.F., 1984, Testing a physically based flood-forecasting model (TOPMODEL) for three U.K. catchments: Journal of Hydrology, v. 69, p. 119–143.

Bokuniewicz, Henry, 1980, Groundwater seepage into Great South Bay, New York: Estuarine and Coastal Marine Science, v. 10, p. 437–444.

Bouwer, H., and Maddock, T., III, 1997, Making sense of the interactions between groundwater and streamflow—Lessons for water masters and adjudicators: Rivers, v. 6, no. 1, p. 19–31.

Buchtele, J., Buchtelová, M., Gallart, F., Latron, J., Llorens, P., Salvany, C., and Herrmann, A., 1998, Rainfall-runoff processes modelling using SACRAMENTO and BROOK models in the Cal Rodó catchment (Pyrenees, Spain), Conference on Catchment Hydrological and Biochemical Processes in Changing Environment: Liblice, Czechoslovakia, p. 13–16.

Burnett, W.C., Kim, G., and Lane-Smith, D., 2001, A continuous monitor for assessment of ^{222}Rn in the coastal ocean: Journal of Radioanalytical and Nuclear Chemistry, v. 249, no. 1, p. 167–172.

Cable, J.E., Bugna, G.C., Burnett, W.C., and Chanton, J.P., 1996, Application of ^{222}Rn and CH$_4$ for assessment of groundwater discharge to the coastal ocean: Limnology and Oceanography, v. 41, no. 6, p. 1347–1353.

Cable, J.E., Burnett, W.C., Chanton, J.P., Corbett, D.R., and Cable, P.H., 1997, Field evaluation of seepage meters in the coastal marine environment: Estuarine, Coastal and Shelf Science, v. 45, no. 3, p. 367–375.

Cambareri, T.C., and Eichner, E.M., 1998, Watershed delineation and ground-water discharge to a coastal embayment: Ground Water, v. 36, no. 4, p. 626–634.

Campbell, C.W., and Keith, A.G., 2001, Karst ground-water hydrologic analyses based on aerial thermography: Hydrological Science and Technology, Special Issue, v. 17, no. 1–4, p. 59–68.

Carr, M.R., and Winter, T.C., 1980, An annotated bibliography of devices developed for direct measurement of seepage: U.S. Geological Survey Open-File Report 80–344, 38 p.

Carter, J.M., Driscoll, D.G., Williamson, J.E., and Lindquist, V.A., 2002, Atlas of water resources in the Black Hills area, South Dakota: U.S. Geological Survey Hydrologic Investigations Atlas HA–747, 120 p.

Cheng, Xiangxue, and Anderson, M.P., 1993, Numerical simulation of ground-water interaction with lakes allowing for fluctuating lake levels: Ground Water, v. 31, no. 6, p. 929–933.

Childress, C.J.O., Sheets, R.A., and Bair, E.S., 1991, Hydrology and water quality near the South Well Field, southern Franklin County, Ohio, with emphasis on the simulation of ground-water flow and transport of Scioto River: U.S. Geological Survey Water-Resources Investigations Report 91–4080, 78 p.

Choi, Jungyill, and Harvey, J.W., 2000, Quantifying time-varying ground-water discharge and recharge in wetlands of the northern Florida Everglades: Wetlands, v. 20, no. 3, p. 500–511.

Committee on Hydrologic Science, National Research Council, 2004, Groundwater fluxes across interfaces: Washington, D.C., The National Academies Press, 85 p.

Conant, Brewster, Jr., 2004, Delineating and quantifying ground-water discharge zones using streambed temperatures: Ground Water, v. 42, no. 2, p. 243–257.

Constantz, James, Stonestrom, David, Stewart, A.E., Niswonger, Rich, and Smith, T.R., 2001, Analysis of streambed temperatures in ephemeral channels to determine streamflow frequency and duration: Water Resources Research, v. 37, no. 2, p. 317–328.

Constantz, James, Thomas, C.L., and Zellweger, G., 1994, Influence of diurnal variations in stream temperature on streamflow loss and groundwater recharge: Water Resources Research, v. 30, no. 12, p. 3253–3264.

Corbett, D.R., Burnett, W.C., Cable, P.H., and Clark, S.B., 1998, A multiple approach to the determination of radon fluxes from sediments: Journal of Radioanalytical and Nuclear Chemistry, v. 236, p. 247–252.

Corbett, D.R., Chanton, Jeffrey, Burnett, William, Dillon, Kevin, Rutkowski, Christine, and Fourqurean, J.W., 1999, Patterns of groundwater discharge into Florida Bay: Limnology and Oceanography, v. 44, no. 4, p. 1045–1055.

Danielopol, D.L., 1984, Ecological investigations on the alluvial sediments of the Danube in the Vienna area—A phreatobiological project: Verhandlungen Internationale Vereiningen Limnologie, v. 22, p. 1755–1761.

Danielopol, D.L., Rouch, R., Pospisil, P., Torreiter, P., and Moeszlacher, F., 1997, Ecotonal animal assemblages: their interest for groundwater studies, in Gibert, J., Mathieu, J., and Fournier, F., eds., Groundwater/surface water ecotones—Biological and hydrological interactions and management options: Cambridge, Cambridge University Press, p. 11–20.

Danskin, W.R., 1998, Evaluation of the hydrologic system and selected water-management alternatives in the Owens Valley, California: U.S. Geological Survey Water-Supply Paper 2370–H, 175 p., 6 pls. in pocket.

Day-Lewis, F.D., White, E.A., Belaval, M., Johnson, C.D., and Lane, J.W., Jr., 2006, Continuous resistivity profiling to delineate submarine ground-water discharge—Examples and limitations: The Leading Edge, v. 25, no. 6, p.724–728.

Dincer, T., 1968, The use of oxygen 18 and deuterium concentrations in the water balance of lakes: Water Resources Research, v. 4, no. 6, p. 1289–1306.

Dokulil, M.T., Teubner, K., and Donabaum, K., 2000, Restoration of a shallow, ground-water fed urban lake using a combination of internal management strategies—A case study: Advances in Limnology, v. 55, p. 271–282.

Donato, M.M., 1998, Surface-water/ground-water relations in the Lemhi River Basin, East-Central Idaho: U.S. Geological Survey Water-Resources Investigations Report 98–4185, 29 p.

Doss, P.K., 1993, Nature of a dynamic water table in a system of non-tidal freshwater coastal wetlands: Journal of Hydrology, v. 141, p. 107–126.

Dumouchelle, D.H., 2001, Evaluation of ground-water/surface-water relations, Chapman Creek, west-central, Ohio, by means of multiple methods: U.S. Geological Survey Water-Resources Investigations Report 2001–4202, 13 p.

Ellins, K.K., Roman-Mas, A., and Lee, R., 1990, Using ^{222}Rn to examine groundwater/surface discharge interaction in the Rio Grande de Manati, Puerto Rico: Journal of Hydrology, v. 115, p. 319–341.

Erickson, D.R., 1981, A study of littoral groundwater seepage at Williams Lake, Minnesota, using seepage meters and wells: Minneapolis, University of Minnesota, M.S. thesis, 135 p.

Federer, C.A., and Lash, D., 1978, BROOK—A hydrologic simulation model for eastern forests: Water Resources Research Center, University of New Hampshire Research Report no. 19, 84 p.

Fellows, C.R., and Brezonik, P.L., 1980, Seepage flow into Florida lakes: Water Resources Bulletin, v. 16, no. 4, p. 635–641.

Fenske, J.P., Leake, S.A., and Prudic, D.E., 1996, Documentation of a computer program (RES1) to simulate leakage from reservoirs using the modular finite-difference groundwater flow model (MODFLOW): U.S. Geological Survey Open-File Report 96–364, 51 p.

Fischer, W.A., Davis, D.A., and Sousa, T.M., 1966, Fresh-water springs of Hawaii from infrared images: U.S. Geological Survey Hydrologic Atlas 218, 1 map.

Fraser, C.J.D., Roulet, N.T., and Lafleur, M., 2001, Groundwater flow pattern in a large peatland: Journal of Hydrology, v. 246, p. 142–154.

Galloway, D.L., Alley, W.M., Barlow, P.M., Reilly, T.E., and Tucci, P., 2003, Evolving issues and practices in managing ground-water resources—Case studies on the role of science: U.S. Geological Survey Circular 1247, 73 p.

Garrett, J.W., Bennett, D.H., Frost, F.O., and Thurow, R.F., 1998, Enhanced incubation success for Kokanee spawning in groundwater upwelling sites in a small Idaho stream: North American Journal of Fisheries Management, v. 18, no. 4, p. 925–930.

Garrison, G.H., Glenn, C.R., and McMurtry, G.M., 2003, Measurement of submarine groundwater discharge in Kahana Bay, O'ahu, Hawai'i: Limnology and Oceanography, v. 48, no. 2, p. 920–928.

Genereux, D.P., and Hemond, H.F., 1990, Naturally occurring radon 222 as a tracer for streamflow generation—Steady-state methodology and field application: Water Resources Research, v. 26, no. 12, p. 3065–3075.

Glennon, Robert, 2002, Water follies—Groundwater pumping and the fate of America's fresh waters: Washington, D.C., Island Press, 314 p.

Goslee, S.C., Brooks, R.P., and Cole, C.A., 1997, Plants as indicators of wetland water source: Plant Ecology, v. 131, p. 199–206.

Harbaugh, A.W., Banta, E.R., Hill, M.C., and McDonald, M.G., 2000, MODFLOW-2000, the U.S. Geological Survey modular ground-water model—User guide to modularization concepts and the ground-water flow process: U.S. Geological Survey Open-File Report 2000–92, 121 p.

Harte, P.T., Flynn, R.J., Kiah, R.G., Severance, T., and Coakley, M.F., 1997, Information on hydrologic and physical properties of water to assess transient hydrology of the Milford-Souhegan glacial-drift aquifer, Milford, New Hampshire: U.S. Geological Survey Open-File Report 97–414, 45 p.

Harvey, F.E., Lee, D.R., Rudolph, D.L., and Frape, S.K., 1997, Locating groundwater discharge in large lakes using bottom sediment electrical conductivity mapping: Water Resources Research, v. 33, no. 11, p. 2609–2615.

Harvey, J.W., and Wagner, B.J., 2000, Quantifying hydrologic interactions between streams and their subsurface hyporheic zones, in Jones, J.B., and Mulholland, P.J., eds., Streams and ground waters: San Diego, Academic Press, p. 3–44.

Healy, R.W., and Ronan, A.D., 1996, Documentation of computer program VS2DH for simulation of energy transport in variably saturated porous media—Modification of the U.S. Geological Survey's computer program VS2DT: U.S. Geological Survey Water-Resources Investigations Report 96–4230, 36 p.

Hemond, H.F., 1983, The nitrogen budget of Thoreau's Bog: Ecology, v. 64, p. 94–109.

Herbert, L.R., and Thomas, B.K., 1992, Seepage study of the Bear River including Cutler Reservoir in Cache Valley, Utah and Idaho: Salt Lake City, Utah Department of Natural Resources, Division of Water Rights, State of Utah Department of Natural Resources Technical Publication no. 105, 18 p.

Hill, M.C., Lennon, G.P., Brown, G.A., Hebson, C.S., and Rheaume, S.J., 1992, Geohydrology of simulation of ground-water flow in the valley-fill deposits in the Ramapo River Valley, New Jersey: U.S. Geological Survey Water-Resources Investigations Report 90–4151, 92 p.

Hussain, N., Church, T.M., and Kim, G., 1999, Use of ^{222}Rn and ^{228}Ra to trace groundwater discharge into the Chesapeake Bay: Marine Chemistry, v. 65, p. 127–134.

Hutson, S.S., Barber, N.L., Kenny, J.F., Linsey, K.S., Lumia, D.S., and Maupin, M.A., 2004, Estimated use of water in the United States in 2000: U.S. Geological Survey Circular 1268, 52 p.

Jacobs, K.L., and Holway, J.M., 2004, Managing for sustainability in an arid climate—Lessons learned from 20 years of groundwater management in Arizona, USA: Hydrogeology Journal, v. 12, no. 1, p. 52–65.

Jaquet, N.G., 1976, Ground-water and surface-water relationships in the glacial province of northern Wisconsin—Snake Lake: Ground Water, v. 14, no. 2, p. 194–199.

Jobson, H.E., and Harbaugh, A.W., 1999, Modifications to the diffusion analogy surface-water flow model (DAFLOW) for coupling to the modular finite-difference ground-water flow model (MODFLOW): U.S. Geological Survey Open-File Report 99-217, 107 p.

Jones, W.K., ed., 1984, Water tracing special issue: National Speleological Society Bulletin, v. 46, no. 2, 48 p.

Johannes, R.E., 1980, The ecological significance of the submarine discharge of groundwater: Marine Ecology—Progress Series, v. 3, p. 365–373.

Johannes, R.E., and Hearn, C.J., 1985, The effect of submarine groundwater discharge on nutrient and salinity regimes in a coastal lagoon off Perth, Western Australia: Estuarine, Coastal and Shelf Science, v. 21, p. 789–800.

Kaleris, V., 1998, Quantifying the exchange rate between groundwater and small streams: Journal of Hydraulic Research, v. 36, no. 6, p. 913–932.

Kilpatrick, F.A., and Cobb, E.D., 1985, Measurement of discharge using tracers: U.S. Geological Survey, Techniques of Water-Resources Investigations, book 3, chap. A–16, 73 p.

Kendall, Carol, and McDonnell, J.J., eds., 1998, Isotope tracers in catchment hydrology: New York, Elsevier Science Publishers, 839 p.

Kendall, Carol, Sklash, M.G., and Bullen, T.K., 1995, Isotope tracers of water and solute sources in catchments, in Trudgill, S.T., ed., Solute modelling in catchment systems: New York, John Wiley & Sons, Ltd., p. 261–303.

Kimball, Briant, 1997, Use of tracer injections and synoptic sampling to measure metal loading from acid mine drainage: U.S. Geological Survey Fact Sheet 245–96, 4 p.

Konrad, C.P., Drost, B.W., and Wagner, B.J., 2003, Hydrogeology of the unconsolidated sediments, water quality, and ground-water/surface-water exchanges in the Methow River Basin, Okanogan County, Washington: U.S. Geological Survey Water-Resources Investigations Report 2003–4244, 137 p.

Kolodny, Yehoshua, Katz, Amitai, Starinsky, Abraham, Moise, Tamar, and Simon, Ehud, 1999, Chemical tracing of salinity sources in Lake Kinneret (Sea of Galilee), Israel: Limnology and Oceanography, v. 44, no. 4, p. 1035–1044.

Krabbenhoft, D.P., and Anderson, M.P., 1986, Use of a numerical ground-water flow model for hypothesis testing: Ground Water, v. 24, p. 49–55.

Krabbenhoft, D.P., Anderson, M.P., and Bowser, C.J., 1990, Estimating groundwater exchange with lakes, 2—Calibration of a three-dimensional, solute transport model to a stable isotope plume: Water Resources Research, v. 26, no. 10, p. 2455–2462.

Krabbenhoft, D.P., Bowser, C.J., Kendall, C., and Gat, J.R., 1994, Use of oxygen-18 and deuterium to assess the hydrology of groundwater-lake systems, in Baker, L.A., ed., Environmental Chemistry of Lakes and Reservoirs: American Chemical Society, Advances in Chemistry Series no. 237, p. 67–90.

Kraemer, T.F., 2005, Radium isotopes in Cayuga Lake, New York—Indicators of inflow and mixing processes: Limnology and Oceanography, v. 50, no. 1, p. 158–168.

Kraemer, T.F., and Genereux, D.P., 1998, Applications of uranium and thorium series radionuclides in catchment hydrology studies, in Kendall, Carol, and McDonnell, J.J., eds., Isotope tracers in catchment hydrology: Amsterdam, Elsevier, p. 679–722.

Krest, J.M., and Harvey, J.W., 2003, Using natural distributions of short-lived radium isotopes to quantify ground water discharge and recharge: Limnology and Oceanography, v. 48, no. 1, p. 290–298.

Krest, J.M., Moore, W.S., Gardner, L.R., and Morris, J.T., 2000, Marsh nutrient export supplied by groundwater discharge—Evidence from radium measurements: Global Biogeochemical Cycles, v. 14, no. 1, p. 167–176.

LaBaugh, J.W., 1985, Uncertainty in phosphorus retention, Williams Fork Reservoir, Colorado: Water Resources Research, v. 21, no. 11, p. 1684–1692.

LaBaugh, J.W., Rosenberry, D.O., and Winter, T.C., 1995, Groundwater contribution to the water and chemical budgets of Williams Lake, Minnesota, 1980–1991: Canadian Journal of Fisheries and Aquatic Science, v. 52, p. 754–767.

LaBaugh, J.W., Winter, T.C., Rosenberry, D.O., Schuster, P.F., Reddy, M.M., and Aiken, G.A., 1997, Hydrological and chemical estimates of the water balance of a closed-basin lake in north-central Minnesota: Water Resources Research, v. 33, no. 12, p. 2799–2812.

Lapham, W.W., 1989, Use of temperature profiles beneath streams to determine rates of vertical ground-water flow and vertical hydraulic conductivity: U.S. Geological Survey Water-Supply Paper 2337, 35 p.

Leavesley, G.H., and Hay, L., 1998, The use of coupled atmospheric and hydrological models for water-resources management in headwater basins, *in* Kovar, K., Tappeiner, U., Peters, N.E., and Craig, R.G., eds., Hydrology, water resources, and ecology in headwaters, Meran, Italy: Meran, Italy IAHS-AISH Publication, p. 259–265.

Leavesley, G.H., Lichty, R.W., Troutman, B.M., and Saindon, L.G., 1983, Precipitation-runoff modeling system—User's manual: U.S. Geological Survey Water-Resources Investigations Report 83–4238, 207 p.

Leavesley, G.H., Markstrom, S.L., Restrepo, P.J., and Viger, R.J., 2002, A modular approach to addressing model design, scale, and parameter estimation issues in distributed hydrological modeling: Hydrological Processes, v. 16, p. 173–187.

Leavesley, G.H., Restrepo, P.J., Markstrom, S.L., Dixon, M., and Stannard, L.G., 1996, The modular modeling system—MMS, User's manual: U.S. Geological Survey Open-File Report 96–151, 142 p.

Lee, D.R., 1977, A device for measuring seepage flux in lakes and estuaries: Limnology and Oceanography, v. 22, no. 1, p. 140–147.

Lee, D.R., 1985, Method for locating sediment anomalies in lakebeds that can be caused by groundwater flow: Journal of Hydrology, v. 79, p. 187–193.

Lee, D.R., and Cherry, J.A., 1978, A field exercise on ground-water flow using seepage meters and minipiezometers: Journal of Geological Education, v. 27, p. 6–20.

Lee, D.R., Cherry, J.A., and Pickens, J.F., 1980, Groundwater transport of a salt tracer through a sandy lakebed: Limnology and Oceanography, v. 25, no. 1, p. 45–61.

Lee, T.M., and Swancar, Amy, 1997, Influence of evaporation, ground water, and uncertainty in the hydrologic budget of Lake Lucerne, a seepage lake in Polk County, Florida: U.S. Geological Survey Water-Supply Paper 2439, 61 p.

Linderfelt, W.R., and Turner, J.V., 2001, Interaction between shallow groundwater, saline surface water and nutrient discharge in a seasonal estuary—The Swan–Canning system: Hydrological Processes, v. 15, p. 2631–2653.

Lindgren, R.J., and Landon, M.K., 1999, Effects of ground-water withdrawals on the Rock River and associated valley aquifer, eastern Rock County, Minnesota: U.S. Geological Survey Water-Resources Investigations Report 99–4157, 103 p.

Lodge, D.M., Krabbenhoft, D.P., and Striegl, R.G., 1989, A positive relationship between groundwater velocity and submersed macrophyte biomass in Sparkling Lake, Wisconsin: Limnology and Oceanography, v. 34, no. 1, p. 235–239.

Loke, M.H., and Lane, J.W., Jr., 2004, Inversion of data from electrical resistivity imaging surveys in water-covered areas: Exploration Geophysics, v. 35, p. 266–271.

Malard, F., Plenet, S., and Gibert, J., 1996, The use of invertebrates in ground water monitoring—A rising research field: Ground Water Monitoring and Remediation, v. 16, no. 2, p. 103–113.

Malcolm, I.A., Soulsby, C., Youngson, A.F., and Petry, J., 2003a, Heterogeneity in ground-water/surface-water interactions in the hyporheic zone of a salmonid spawning stream: Hydrological Processes, v. 17, p. 601–617.

Malcolm, I.A., Youngson, A.F., and Soulsby, Chris, 2003b, Survival of salmonid eggs in a degraded gravel-bed stream—Effects of ground-water/surface-water interactions: Hydrobiologia, v. 444, p. 303–316.

Manheim, F.T., Krantz, D.E., and Bratton, J.F., 2004, Studying ground water under Delmarva coastal bays using electrical resistivity: Ground Water, v. 42, no. 7, p. 1052–1068.

McBride, M.S., and Pfannkuch, H.O., 1975, The distribution of seepage within lakebeds: U.S. Geological Survey Journal of Research, v. 3, no. 5, p. 505–512.

McCarthy, K.A., McFarland, W.D., Wilkinson, W.D., and White, L.D., 1992, The dynamic relationship between ground water and the Columbia River—Using deuterium and oxygen-18 as tracers: Journal of Hydrology, v. 135, p. 1–12.

McLeod, R.S., 1980, The effects of using ground water to maintain water levels of Cedar Lake, Wisconsin: U.S. Geological Survey Water-Resources Investigations Report 80–23, 35 p.

Merritt, M.L., and Konikow, L.F., 2000, Documentation of a computer program to simulate lake-aquifer interaction using the MODFLOW Ground-Water Flow Model and the MOC3D Solute-Transport Model: U.S. Geological Survey Water-Resources Investigations Report 00–4167, 146 p.

Metz, P.A., and Sacks, L.A., 2002, Comparison of the hydrogeology and water quality of a ground-water augmented lake with two non-augmented lakes in northwest Hillsborough County, Florida: U.S. Geological Survey Water-Resources Investigations Report 02–4032, 74 p.

Meyboom, P., 1966, Unsteady groundwater flow near a willow ring in hummocky moraine: Journal of Hydrology, v. 4, p. 38–62.

Meyboom, P., 1967, Mass-transfer studies to determine the groundwater regime of permanent lakes in hummocky moraine of western Canada: Journal of Hydrology, v. 5, p. 117–142.

Miller, R.L., Kraemer, T.F., and McPherson, B.F., 1990, Radium and radon in Charlotte Harbor Estuary, Florida: Estuarine, Coastal and Shelf Science, v. 31, p. 439–457.

Mitchell-Bruker, S., and Haitjema, H.M., 1996, Modeling steady state conjunctive groundwater and surface water flow with analytic elements: Water Resources Research, v. 32, no. 9, p. 2725–2732.

Moore, W.S., 1996, Large groundwater inputs to coastal waters revealed by ^{228}Ra enrichments: Nature, v. 380, p. 612–614.

Moore, W.S., 1999, The subterranean estuary—A reaction zone of ground water and sea water: Marine Geochemistry, v. 65, p. 111–125.

Moore, W.S., 2000, Determining coastal mixing rates using radium isotopes: Continental Shelf Research, v. 20, p. 1993–2007.

Mull, D.S., Liebermann, T.D., Smoot, J.L., and Woosley, L.H., Jr., 1988, Application of dye-tracing techniques for determining solute transport characteristics of ground water in karst terranes: U.S. Environmental Protection Agency Report 904/6–88–001.

Niswonger, R.G., and Prudic, D.E., 2005, Documentation of the streamflow-routing (SFR2) package to include unsaturated flow beneath streams—A modification to SFR1: U.S. Geological Survey Techniques and Methods 6–A13, 48 p.

Niswonger, R.G., Markstrom, R.L., Regan, R.S., Prudic, D.E., Pohll, Greg, and Viger, R.J., 2006, Modeling ground-water/surface-water interactions with GSFLOW, a new USGS model: Golden, Colo., Proceedings, MODFLOW and More 2006, Managing Ground Water Systems, May 21–24, 2006, p. 99–103.

Oberg, K.A., Morlock, S.E., and Caldwell, W.S., 2005, Quality-assurance plan for discharge measurements using acoustic Doppler current profilers: U.S. Geological Survey Scientific Investigations Report 2005–5183, 44 p.

Orghidan, T., 1959, Ein neuer Lebensraum des unteriridischen Wassers: Der hyporheische Biotop: Archiv für Hydrobiologie, v. 55, p. 392–414.

Owen-Joyce, S.J., Wilson, R.P., Carpenter, M.C., and Fink, J.B., 2000, Method to identify wells that yield water that will be replaced by water from the Colorado River downstream from Laguna Dam in Arizona and California: U.S. Geological Survey Water-Resources Investigations Report 00–4085, 31 p. 19 maps.

Pfannkuch, H.O., and Winter, T.C., 1984, Effect of anisotropy and groundwater system geometry on seepage through lakebeds, 1, Analog and dimensional analysis: Journal of Hydrology, v. 75, p. 213–237.

Pluhowski, E.J., 1972, Hydrologic interpretations based on infrared imagery of Long Island, New York, Contributions to the hydrology of the United States: U.S. Geological Survey Water-Supply Paper 2009–B, 20 p.

Power, G., Brown, R.S., and Imhof, J.G., 1999, Groundwater and fish—Insights from northern North America: Hydrological Processes, v. 13, p. 401–422.

Prudic, D.E., 1989, Documentation of a computer program to simulate stream-aquifer relations using the modular finite-difference ground-water flow model: U.S. Geological Survey Open-File Report 88–729, 113 p.

Prudic, D.E., Konikow, L.F., and Banta, E.R., 2004, A new streamflow-routing (SFR1) package to simulate stream-aquifer interaction with MODFLOW-2000: U.S. Geological Survey Open-File Report 2004–1042, 95 p.

Puckett, L.J., Cowdery, T.K., McMahon, P.B., Tornes, L.H., and Stoner, J.D., 2002, Using chemical, hydrologic, and age dating analysis to delineate redox processes and flow paths in the riparian zone of a glacial outwash aquifer-stream system: Water Resources Research, v. 38, no. 8, p. 9–1 to 9–20.

Rantz, S.E., and others, 1982a, Measurement and computation of streamflow—Volume 1, Measurement of stage and discharge: U.S. Geological Survey Water-Supply Paper 2175, 284 p.

Rantz, S.E., and others, 1982b, Measurement and computation of streamflow—Volume 2, Computation of discharge: U.S. Geological Survey Water-Supply Paper 2175, 631 p.

Rimmer, Alon, and Gideon, Gal, 2003, Estimating the saline springs component in the solute and water balance of Lake Kinneret, Israel: Journal of Hydrology, v. 284, p. 228–243.

Robinove, C.J., 1965, Infrared photography and imagery in water resources research: Journal of the American Water Works Association, v. 57, pt. 2, p. 834–840.

Robinove, C.J., and Anderson, D.G., 1969, Some guidelines for remote sensing in hydrology: Water Resources Bulletin, v. 5, no. 2, p. 10–19.

Rosenberry, D.O., 1990, Inexpensive groundwater monitoring methods for determining hydrologic budgets of lakes and wetlands, in National Conference on Enhancing the States' Lake and Wetland Management Programs: U.S. Environmental Protection Agency, North American Lake Management Society, p. 123–131.

Rosenberry, D.O., 2005, Integrating seepage heterogeneity with the use of ganged seepage meters: Limnology and Oceanography/Methods, v. 3, p. 131–142.

Rosenberry, D.O., Bukaveckas, P.A., Buso, D.C., Likens, G.E., Shapiro, A.M., and Winter, T.C., 1999, Migration of road salt to a small New Hampshire lake: Water Air and Soil Pollution, v. 109, p. 179–206.

Rosenberry, D.O., Striegl, R.G., and Hudson, D.C., 2000, Plants as indicators of focused ground water discharge to a northern Minnesota lake: Ground Water, v. 38, no. 2, p. 296–303.

Rosenberry, D.O., and Winter, T.C., 1997, Dynamics of water-table fluctuations in an upland between two prairie-pothole wetlands in North Dakota: Journal of Hydrology, v. 191, p. 266–289.

Rundquist, D., Murray, G., and Queen, L., 1985, Airborne thermal mapping of a "flow-through" lake in the Nebraska sandhills: Water Resources Research, v. 21, no. 6, p. 989–994.

Rutledge, A.T., 1992, Methods of using streamflow records for estimating total and effective recharge in the Appalachian Valley and Ridge, Piedmont, and Blue Ridge physiographic provinces: American Water Resource Association Monograph Series no. 17, p. 59–73.

Rutledge, A.T., 1998, Computer programs for describing the recession of ground-water discharge and for estimating mean ground-water recharge and discharge from streamflow data—Update: U.S. Geological Survey Water-Resources Investigations Report 98–4148, 44 p.

Rutledge, A.T., 2000, Considerations for use of the RORA program to estimate ground-water recharge from streamflow records: U.S. Geological Survey Open-File Report 2000–156, 52 p.

Sacks, L.A., 2002, Estimating ground-water inflow to lakes in central Florida using the isotope mass-balance approach: U.S. Geological Survey Water-Resources Investigations Report 02–4192, 59 p.

Sacks, L.A., Swancar, Amy, and Lee, T.M., 1998, Estimating ground-water exchange with lakes using water-budget and chemical mass-balance approaches for ten lakes in ridge areas of Polk and Highlands Counties, Florida: U.S. Geological Survey Water-Resources Investigations Report 98–4133, 52 p.

Schindler, D.W., Newbury, R.W., Beaty, K.G., and Campbell, P., 1976, Natural water and chemical budgets for a small Precambrian lake basin in central Canada: Journal of the Fisheries Research Board of Canada, v. 33, no. 11, p. 2526–2543.

Sheets, R.A., Darner, R.A., and Whitteberry, B.L., 2002, Lag times of bank filtration at a well field, Cincinnati, Ohio, USA: Journal of Hydrology, v. 266, p. 162–174.

Simmons, G.M., Jr., 1992, Importance of submarine groundwater discharge (SGWD) and seawater cycling to material flux across sediment/water interfaces in marine environments: Marine Ecology Progress Series, v. 81, p. 173–184.

Simonds, F.W., and Sinclair, K.A., 2002, Surface-water/ground-water interactions along the lower Dungeness River and vertical hydraulic conductivity of streambed sediments, Clallam County, Washington, September 1999–July 2001: U.S. Geological Survey Water-Resources Investigations Report 02–4161, 60 p.

Simonds, F.W., Longpre, C.I., and Justin, G.B., 2004, Ground-water system in the Chimacum Creek Basin and surface-water/ground-water interaction in Chimacum and Tarboo Creeks and the Big and Little Quilcene Rivers, Eastern Jefferson County, Washington: U.S. Geological Survey Scientific Investigations Report 2004–5058, 60 p., one plate.

Smart, P.L., and Laidlaw, I.M.S., 1977, An evaluation of some fluorescent dyes for water tracing: Water Resources Research, v. 13, p. 15–33.

Sophocleous, Marios, 2002, Interactions between groundwater and surface water—The state of the science: Hydrogeology Journal, v. 10, p. 52–67.

Sophocleous, M.A., Koelliker, J.K., Govindaraju, R.S., Birdie, T., Ramireddygari, S.R., and Perkins, S.P., 1999, Integrated numerical modeling for basin-wide water management—The case of the Rattlesnake Creek basin in south-central Kansas: Journal of Hydrology, v. 214, p. 179–196.

Sophocleous, Marios, and Perkins, S.P., 2000, Methodology and application of combined watershed and ground-water models in Kansas: Journal of Hydrology, v. 236, p. 185–201.

Steele, G.V., and Verstraeten, I.M., 1999, Effects of pumping collector wells on river—Aquifer interactions at Platte River Island near Ashland, Nebraska, 1998: U.S. Geological Survey Water-Resources Investigations Report 99–4161, 6 p.

Stewart, J.W., and Hughes, G.H., 1974, Hydrologic consequences of using groundwater to maintain lake levels affected by water wells near Tampa, Florida: Tallahassee, Florida, Bureau of Geology Report of Investigation, no. 74, 41 p.

Stonestrom, D.A., and Constanz, Jim, eds., 2003, Heat as a tool for studying the movement of ground water near streams: U.S. Geological Survey Circular 1260, 105 p. (http://water.usgs.gov/pubs/circ/2003/circ1260/pdf/Circ1260.pdf)

Strobel, M.L., Sawyer, J.F., and Rahn, P.H., 2000, Field trip road log—Hydrogeology of the central Black Hills of South Dakota, in Strobel, M.L., and others, eds., Hydrology of the Black Hills, Proceedings of the 1999 Conference on the Hydrology of the Black Hills: Rapid City, South Dakota School of Mines and Technology Bulletin no. 20, p. 239–245.

Stromberg, J.C., Tiller, R., and Richter, B., 1996, Effects of groundwater decline on riparian vegetation of semiarid regions—The San Pedro, Arizona: Ecological Applications, v. 6, no. 1, p. 113–131.

Swain, E.D., and Wexler, E.J., 1996, A coupled surface-water and ground-water flow model (MODBRANCH) for simulation of stream-aquifer interaction: U.S. Geological Survey Techniques of Water-Resources Investigations, book 6, chap. A6, 125 p.

Squillace, P.J., Thurman, E.M., and Furlong, E.T., 1993, Groundwater as a nonpoint source of atrazine and deethylatrazine in a river during base flow conditions: Water Resources Research, v. 29, no. 6, p. 1719–1729.

Taniguchi, M., 1993, Evaluation of vertical groundwater fluxes and thermal properties of aquifers based on transient temperature-depth profiles: Water Resources Research, v. 29, no. 7, p. 2021–2026.

Taniguchi, Makato, Burnett, W.C., Cable, J.E., and Turner, J.V., 2002, Investigation of submarine groundwater discharge: Hydrological Processes, v. 16, p. 2115–2129.

Taniguchi, M., Ishitobi, T., and Shimada, J., 2006, Dynamics of submarine groundwater discharge and freshwater-seawater interface: Journal of Geophysical Research, v. 111, C01008, doi:10.1029/2005JC002924.

Taylor, J.I., and Stingelin, R.W., 1969, Infrared imaging for water resources studies: Journal of the Hydraulics Division, Proceedings of the American Society of Civil Engineers, v. 95, no. 1, p. 175–189.

Thoms, R.B., Johnson, R.L., and Healy, R.W., 2006, User's guide to the Variably Saturated Flow (VSF) Process for MODFLOW: U.S. Geological Survey Techniques and Methods 6–A18, 58 p.

Vaccaro, J.J., and Maloy, K.J., 2006, A thermal profile method to identify potential ground-water discharge areas and preferred salmonid habitats for long river reaches: U.S. Geological Survey Scientific Investigations Report 2006–5136, 16 p. (http://pubs.water.usgs.gov/sir2006-5136/)

Valiela, Ivan, Costa, Joseph, Foreman, Kenneth, Teal, J.M., Howes, Brian, and Aubrey, David, 1990, Transport of groundwater-borne nutrients from watersheds and their effects on coastal waters: Biogeochemistry, v. 10, p. 177–197.

Vanek, Vladimir, and Lee, D.R., 1991, Mapping submarine groundwater discharge areas—An example from Laholm Bay, southwest Sweden: Limnology and Oceanography, v. 36, no. 6, p. 1250–1262.

Wentz, D.A., Rose, W.J., and Webster, K.E., 1995, Long-term hydrologic and biogeochemical responses of a soft water seepage lake in north central Wisconsin: Water Resources Research, v. 31, no. 1, p. 199–212.

Wetzel, R.G., 1999, Plants and water in and adjacent to lakes, in Baird, A.J., and Wilby, R.L., eds., Eco-Hydrology: New York, Routledge, p. 269–299.

Winter, T.C., 1986, Effect of ground-water recharge on configuration of the water table beneath sand dunes and on seepage in lakes in the Sandhills of Nebraska, U.S.A.: Journal of Hydrology, v. 86, p. 221–237.

Winter, T.C., 1999, Relation of streams, lakes, and wetlands to groundwater flow systems: Hydrogeology Journal, v. 7, p. 28–45.

Winter, T.C., 2001, The concept of hydrologic landscapes: Journal of the American Water Resources Association, v. 37, no. 2, p. 335–349.

Winter, T.C., Harvey, J.W., Franke, O.L., and Alley, W.M., 1998, Ground water and surface water a single resource: U.S. Geological Survey Circular 1139, 79 p.

Winter, T.C., LaBaugh, J.W., and Rosenberry, D.O., 1988, The design and use of a hydraulic potentiomanometer for direct measurement of differences in hydraulic head between groundwater and surface water: Limnology and Oceanography, v. 33, no. 5, p. 1209–1214.

Winter, T.C., and Rosenberry, D.O., 1995, The interaction of ground water with prairie pothole wetlands in the Cottonwood Lake area, east-central North Dakota, 1979–1990: Wetlands, v. 15, no. 3, p. 193–211.

Woessner, W.W., 1998, Changing views of stream-ground-water interaction, in Van Brahana, J., Eckstein, Y., Ongley, L.K., Schneider, R., and Moore, J.E., eds., Gambling with groundwater—Physical, chemical, and biological aspects of aquifer-stream relations: Las Vegas, Nev., Proceedings of the Joint Meeting of the XXVIII Congress of the International Association of Hydrogeologists and the annual meeting of the American Institute of Hydrologists, Las Vegas, Nevada, September 28 to October 2, 1998, p. 1–6.

Woessner, W.W., 2000, Stream and fluvial plain ground water interactions—Rescaling hydrogeologic thought: Ground Water, v. 38, no. 3, p. 423–429.

Woessner, W.W., and Sullivan, K.E., 1984, Results of seepage meter and mini-piezometer study, Lake Mead, Nevada: Ground Water, v. 22, no. 5, p. 561–568.

Zarriello, P.J., and Reis, K.G., III, 2000, A precipitation-runoff model for the analysis of the effects of water withdrawals on streamflow, Ipswich River Basin, Massachusetts: U.S. Geological Survey Water-Resources Investigations Report 00–4029, 99 p.

Zekster, I.S., 1996, Groundwater discharge into lakes—A review of recent studies with particular regard to large saline lakes in central Asia: International Journal of Salt Lake Research, v. 4, p. 233–249.

Use of Monitoring Wells, Portable Piezometers, and Seepage Meters to Quantify Flow Between Surface Water and Ground Water

By Donald O. Rosenberry, James W. LaBaugh, and Randall J. Hunt

Chapter 2 of
Field Techniques for Estimating Water Fluxes Between Surface Water and Ground Water
Edited by Donald O. Rosenberry and James W. LaBaugh

Techniques and Methods Chapter 4–D2

U.S. Department of the Interior
U.S. Geological Survey

Contents

Figures

Tables

Chapter 2
Use of Monitoring Wells, Portable Piezometers, and Seepage Meters to Quantify Flow Between Surface Water and Ground Water

By Donald O. Rosenberry, James W. LaBaugh, and Randall J. Hunt

Introduction

This chapter describes three of the most commonly used methods to either calculate or directly measure flow of water between surface-water bodies and the ground-water domain. The first method involves measurement of water levels in a network of wells in combination with measurement of the stage of the surface-water body to calculate gradients and then water flow. The second method involves the use of portable piezometers (wells) or hydraulic potentiomanometers to measure gradients. In the third method, seepage meters are used to measure directly flow across the sediment-water interface at the bottom of the surface-water body. Factors that affect measurement scale, accuracy, sources of error in using each of the methods, common problems and mistakes in applying the methods, and conditions under which each method is well- or ill-suited also are described.

Water-Level Measurements and Flow-Net Analysis

The flow-net analysis method, often called the "Darcy approach," is probably the most frequently used method for quantifying flow between ground water and surface water, especially on a whole-lake or watershed scale. In this method, a combination of measurements of water levels in near-shore water-table wells and measurements of water stage of adjacent surface-water bodies are used to calculate water-table gradients between the wells and the surface-water body. Two approaches commonly are used. One approach segments the shoreline of the surface-water body, depending on the number and location of nearby wells. The second approach generates equipotential lines based on hydraulic-head and surface-water stage data, and uses flow-net analysis to calculate flows to and from the surface-water body. Both methods are described in the following section.

Values of hydraulic conductivity (K), which also are needed to quantify flow, commonly are determined from single-well slug tests conducted in the same wells in which water levels are measured to calculate hydraulic gradients (although a multiple-well aquifer test that encompasses a large volume of aquifer often provides a better indication of hydraulic conductivity appropriate to a lake or watershed scale). Spatial resolution of hydraulic-head gradients and flow between ground water and surface water is directly related to geologic heterogeneity; the greater the heterogeneity of an aquifer, the larger the number of data points (wells) that will be needed to accurately determine hydraulic conditions. Heterogeneity often is difficult to determine in practice, and in many instances, ranges of reasonable values for K are used to estimate the range of flows.

Segmented Approach

In this approach, the shoreline of a surface-water body is divided into segments, with the number of segments depending on the location and number of nearby monitoring wells (fig. 1). For each shoreline segment and associated well, hydraulic conductivity and the gradient between the well and the surface-water body are applied to the entire segment. The length of the shoreline segment, m, is multiplied by the effective thickness of the aquifer, b, to determine the area, A, of a vertical plane at the shoreline through which water passes to either enter or leave the surface-water body (fig. 2). The Darcy equation commonly is used to calculate the flow of water that passes through the vertical plane associated with each segment:

$$Q = KA \frac{(h_1 - h_2)}{L}. \qquad (1)$$

where

Q is flow through a vertical plane that extends beneath the shoreline of a surface-water body (L^3/T);

K is horizontal hydraulic conductivity (L/T);

A is the area of the plane through which all water must pass to either enter or leave the surface-water body, depending on the direction of flow [shoreline length (m) $\times$ effective thickness of the aquifer (b)] (L^2);

h_1 is hydraulic head in the well of interest (L);

h_2 is surface-water stage (L);

and

L is distance from the well to the shoreline (L).

Flows to or from the surface-water body are summed to calculate net flow for the entire surface-water body. This method assumes that:

1. All water that exchanges with a surface-water body passes horizontally through a vertical plane positioned at the shoreline that extends to a finite depth (b) beneath the surface of the surface-water body. At depths greater than b, ground water flows beneath the surface-water body and does not exchange with the surface-water body;

2. The direction of water flow is perpendicular to the shoreline as flow enters or leaves the surface-water body;

3. The gradient (water-table slope) between the well and the surface-water body is uniform; and

4. The aquifer is homogeneous and isotropic within the segment.

Although the Darcy equation is most commonly used in calculating flows between ground water and surface water, its assumption of a constant aquifer thickness is violated where the water table slopes in the vicinity of a surface-water body.

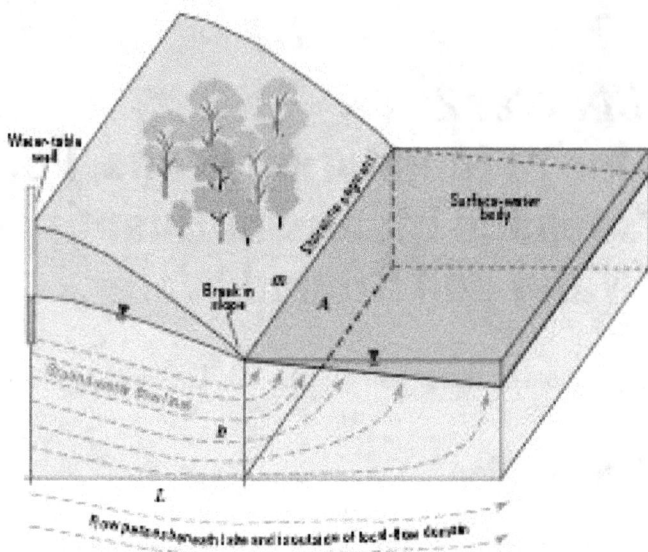

Figure 2.   Typical hydraulic conditions in the vicinity of the shoreline of a surface-water body. (Artwork by Donald O. Rosenberry.)

In these near-shore, unconfined aquifer settings, the use of the Dupuit equation may be more appropriate because it allows the sloping ground-water table to be the upper boundary of the ground-water domain. The Dupuit equation can be written as:

$$Q = Km\frac{(h_1^2 - h_2^2)}{2L}, \qquad (2)$$

where

h_1 = aquifer thickness at the well,

and

h_2 = aquifer thickness at the edge of the surface-water body.

The Dupuit equation assumptions are:

1. The sediments are homogeneous and isotropic;

2. Flow in the aquifer is parallel to the slope of the water table; and

3. For small water-table gradients, ground-water flow lines (also called streamlines) are horizontal.

These assumptions require that equipotential lines (lines of equal hydraulic head) are perpendicular to the ground-water flow lines and are vertical.

As indicated in figure 2, near the shoreline, where water-table gradients typically steepen, these assumptions are violated to some degree. If flow is parallel to the water table, and the water-table gradient is sufficiently steep, then flow in the ground-water system obviously cannot be horizontal. Errors that result from violating these assumptions typically are minor relative to the uncertainty in determining K.

Another source of uncertainty in applying the Dupuit equation is the determination of h_1 and h_2. As with use of the Darcy equation, h_1 and h_2 should include only ground-water

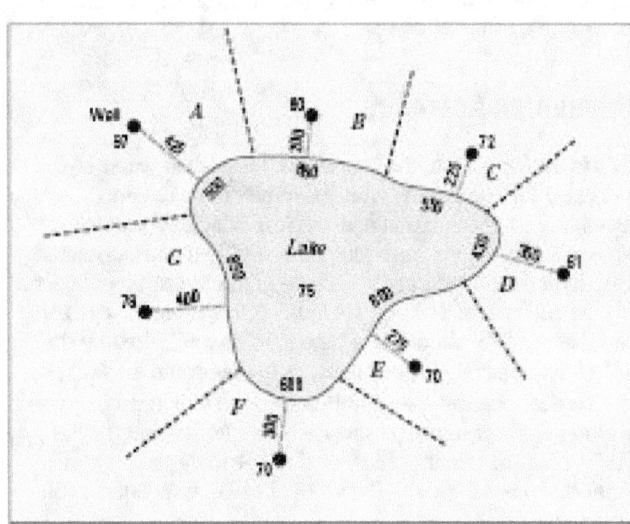

| 0 | 200 | 400 | 600 | 800 | 1,000 METERS |
| 0 | | 1,000 | 2,000 | 3,000 FEET | |

EXPLANATION

$\overset{\bullet}{{}_{50}}$ Well location and ground-water level

- - - -   Watershed segment boundary

A Watershed segment designation

⬓   Distance from well to shoreline, in meters

⬓   Shoreline length per watershed segment, in meters

Figure 1.   A hypothetical lake segmented based on positioning of near-shore water-table wells. Values are hydraulic head and surface-water stage.

flow lines that intersect the surface-water body and exclude those flow lines that pass beneath the surface-water body, in which case $h_2 = b$ as shown in figure 2. This is especially important in cases where a lake, stream, or wetland occupies only the shallow, surficial part of a thick aquifer. As discussed later, h_2 or b often is one of the more difficult parameters to determine.

Although the use of the Dupuit equation is more appropriate for unconfined aquifer settings, the error that results from using the Darcy equation instead of the Dupuit equation commonly is small relative to the uncertainty in determining K. For a small water-table gradient, the errors are very small, and errors are small even for a relatively large water-table gradient. For example, assuming a large water-table gradient of 0.1 and the following values for a 1-meter shoreline reach ($h_1 = 60$ meters, $h_2 = 50$ meters, $L = 100$ meters, $K = 10$ meters per day), the Dupuit flow (Q) = 55 cubic meters per day, and the Darcy flow (Q) = 50 cubic meters per day.

An example of the use of the Darcy approach to calculate flows to and from a surface-water body using values obtained from figure 1 is shown in table 1. The example assumes that K is 30 meters per day and is uniform throughout the watershed, and that b is 20 meters. The method assumes that the hinge lines, the locations where flow direction changes from flow into the lake to flow out of the lake, occur at the ends of the adjacent shoreline segments where a change in flow direction is indicated. Considering the uncertainty associated with positioning of the hinge lines, the difference between total flow into the lake and total flow out of the lake is remarkably small in this example.

Flow-Net Analysis

The flow-net analysis is a graphical method for solving steady-state two-dimensional ground-water flow. The analysis uses the Darcy equation to solve for flow, the distribution of which is dependent on the flow net that is generated manually or with computer software. The method assumes that steady-state flow is two-dimensional (either in plan view, as applied here, or along a cross section), the aquifer is homogeneous and isotropic, and that b (the effective thickness of the aquifer) is known. Rules regarding construction of flow nets are described in Fetter (2000) and in other hydrogeology texts (for example, Davis and DeWiest, 1991). A detailed analysis of the method is provided in Cedergren (1997). In brief, the flow net consists of equipotential lines (lines of equal hydraulic head) and flow lines (also called streamlines). Equipotential lines are drawn on the basis of hydraulic head in the wells and the stage of the surface-water body. They intersect no-flow boundaries at right angles. Assuming the porous medium is homogeneous and isotropic, flow lines are drawn perpendicular to the equipotential lines. A sufficient number of flow lines are drawn so that the resulting rectilinear shapes form approximate squares. The areas between the flow lines are called streamtubes. The intervals between equipotential lines are termed "head drops." Once the flow net is constructed, a form of the Darcy equation is used to approximate flow to or from the surface-water body:

$$Q = \frac{MKbH}{n}. \qquad (3)$$

where

M	=	the number of streamtubes across a flow net,
H	=	total head drop across the area of interest (L),
n	=	number of equipotential head drops over the area of interest, and Q, K, and b are as defined previously.

An example of the flow-net approach is shown in figure 3. The flow net is created using the same hypothetical setting shown in figure 1. The flow domain has been rotated

Table 1.   Data for calculating flows to and from the lake shown in figure 1, and total flow per segment (Q_t) using the segmented Darcy approach.

[m/d, meters per day; m, meter; m³/d, cubic meters per day]

Watershed segment	Horizontal hydraulic conductivity (K) (m/d)	Effective thickness of the aquifer (b) (m)	Hydraulic head in well—surface-water stage (h_1–h_2) (m)	Distance from the well to the shoreline (L) (m)	Length of shoreline segment (m)	Water flow (Q) (m³/d)
A	30	20	22	425	500	15,529
B	30	20	5	200	650	9,750
C	30	20	–3	225	550	–4,400
D	30	20	–14	350	430	–10,320
E	30	20	–5	275	800	–8,727
F	30	20	–5	300	600	–6,000
G	30	20	3	400	850	3,825

Total flow into lake = 29,104 cubic meters per day.

Total flow out of lake = 29,447 cubic meters per day.

so that equipotential lines are approximately perpendicular to the no-flow boundaries on the top and bottom of the
figure, and streamlines are approximately perpendicular to
constant-head boundaries to the left and right of the figure.
The equipotential lines represent hydraulic-head intervals of
10 meters. The total flow of water that exchanges with the lake
is apportioned into seven streamtubes. Using the same values
for K, b, hydraulic head, and lake stage as for the segmented
Darcy method, and values of 7, 35, and 3.5 for M, H, and n,
respectively, the total Q into the lake is 42,000 cubic meters
per day. Total Q out of the lake, based on the same values
as for the segmented Darcy method of 7, 15, and 1.5 for M,
H, and n, respectively, also is 42,000 cubic meters per day.
These values are substantially larger (44 percent) than total
flows into and out of the lake calculated by the segmented
Darcy method.

A comparison of results from the two methods indicates
the relative accuracy of these methods. Substantial errors
can result with the segmented Darcy method if conditions
along each shoreline segments are not uniform. For example,
determination of flow across the curving segment on the
northwest side of the lake assumes that an arc of hydraulic
head 22 meters higher than the lake surface exists a distance
of 425 meters from shore along the entire shoreline segment.
Common sense and the flow-net analysis (fig. 3) indicate that

this is a poor assumption. Also, incorrect placement of the
hinge-line location can result in shoreline segments drawn
adjacent to the hinge line that poorly represent the actual local
flow into and out of the surface-water feature. Figure 4 shows
the flow lines drawn in figure 3 in addition to the shoreline
segments indicated in figure 1. Positioning of hinge lines in
figure 4 is based on the flow-net analysis. If the segmented
Darcy method was used to place hinge lines, they would be
located at the boundaries between segments B and C, and
between segments F and G. Fortunately, a misplacement of
the hinge line commonly does not result in substantial error
because flow across the sediment-water interface commonly
is small where ground-water flow is primarily parallel to
the shoreline.

The flow-net analysis method provides a simple, initial
estimate of the exchange of water between a surface-water
body and ground water. The accuracy of the method depends
on the degree to which the simplifying assumptions are met
in the setting being analyzed and on how well the mesh
is drawn. Errors can be minimized by ensuring that areas
contained by the streamtubes and equipotential lines form
approximate squares. Cedergren (1997) provides additional
information for minimizing mesh-related errors. Uncertainties associated with accurate representation of K commonly
are significantly larger than errors associated with improperly
constructed flow-net meshes. With a larger number of wells,
equipotential lines can be placed more precisely and a finer
grid then can be generated. Accuracy also depends on how
the flow net is interpreted. For example, in the setting shown
in figure 3, streamtubes that partly intersect the lake were
ignored. Those streamtubes might instead have been considered as half streamtubes, in which case the total flow into and
out of the lake would have been larger. Alternately, streamtubes 1 and 7 could have been drawn so as to bypass the lake,
in which case only five streamtubes would intersect the lake. If
the number of streamtubes was five instead of seven, the flow-
net-derived fluxes to and from the lake (30,000 cubic meters
per day) would be nearly identical to the segmented Darcy-
generated fluxes.

The domain shown in figure 3, although rotated to be
aligned with flow lines and equipotential lines, was drawn
with the same dimensions as the domain shown in figure 1.
One could argue, however, that the domain should have been
made larger because many of the flow lines and equipotential
lines do not intersect the boundaries at right angles.

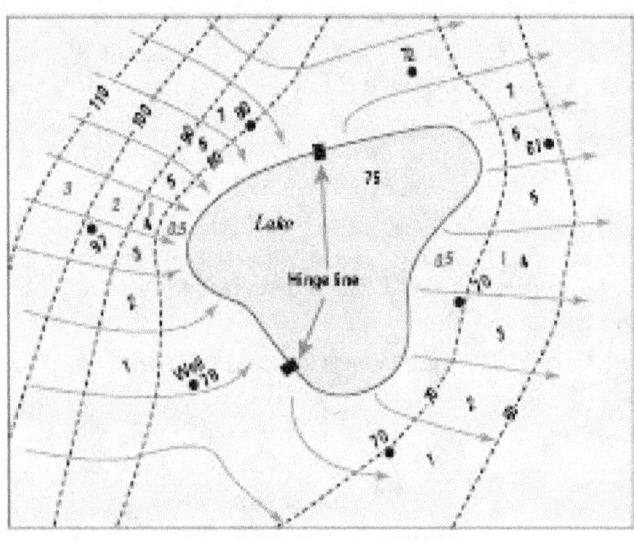

0   200 400 600 800 1,000 METERS

0   1,000   2,000   3,000 FEET

EXPLANATION

- 75 - -   Equipotential line
              Flow line
57   Well location and ground-water level
1   Streamtube number
0.5   Equipotential head drop

Figure 3.   A flow net generated to indicate flow of water to and
from a hypothetical lake. Ground-water flow direction is indicated
by flow lines (blue lines), and lines of equal hydraulic head
(equipotential lines) are shown with dashed lines. Values shown
are hydraulic head in the wells and surface-water stage.

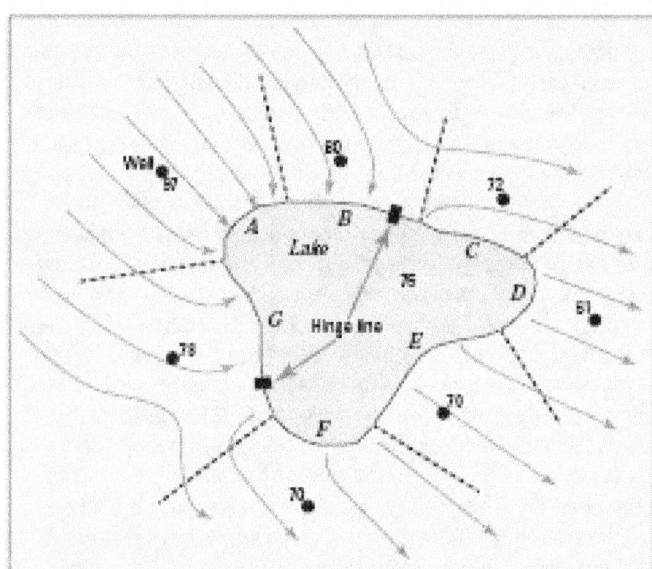

Figure 4.   Conceptualization of flow based on flow-net analysis and segmented Darcy fluxes. The position of the hinge line changes depending on the method of analysis used.

Sources of Error

Sources of error in applying the segmented-shore or flow-net-analysis approach to the determination of the exchanges between a surface-water body and ground water, in addition to errors in interpretation presented above, include:

1. Inadequate physical characterization of conditions or properties that affect flow,
2. Measurement error,
3. Improperly constructed wells,
4. Improperly maintained wells,
5. Unstable wells and stage gage, and
6. Violation of underlying assumptions.

Each item is discussed in detail below.

Inadequate Physical Characterization

In the examples given above, horizontal hydraulic conductivity (K) was assumed to be uniform across the entire watershed. This is a poor assumption because erosional and depositional conditions near the shoreline commonly are different than for the larger watershed. Where lower-K sediments line lakes or wetlands, K within a meter of the sediment-water interface can be the dominant control on flow (Rosenberry, 2000). This is especially well documented in fluvial settings (for example, Brunke, 1999; Hiscock and Grischek, 2002; Schubert, 2002; Sheets and others, 2002; Fleckenstein and others, 2006). It usually is beneficial to install additional wells near the shoreline of the surface-water body to gain a better understanding of the distribution of hydraulic head and of the spatial variability in K.

A slug test can be expected to provide only an approximate estimate of the actual K that controls flow between ground water and surface water. First, slug tests measure horizontal K, but aquifers commonly are anisotropic; vertical K typically is smaller, sometimes orders of magnitude smaller, than horizontal K. Second, measurements of K are to some extent scale dependent and single-well slug tests may provide values that are too small to be representative of the larger scale flow in the aquifer. Rovey and Cherkauer (1995) found that K of a carbonate aquifer in Wisconsin increases linearly with the scale of the measurement up to a radius of influence of between 20 and 220 meters, after which point K was constant with increasing radius. Schulze-Makuch and others (1999) indicated that scale dependence of K depends on the hydraulic properties of an aquifer. They reported that K is relatively insensitive to scale for homogeneous aquifers but increases by half an order of magnitude for every order of magnitude increase in spatial scale of heterogeneous aquifers. Unless the well is installed in the lake, the approaches outlined herein do not attempt to quantify exchange between ground water and surface water at the surface-water feature itself. Rather, they estimate the flow into and out of the ground-water system near the surface-water feature, at the locations of the monitoring wells and assume that water that crosses the vertical plane at the shoreline must either originate from or flow into the lake.

Determination of the effective thickness of the aquifer (b) through which water flows to interact with a surface-water body also can be difficult. Investigators may resort to hypothetical flow modeling or to tracers to address this issue. Siegel and Winter (1980) and Krabbenhoft and Anderson (1986) used finite-difference ground-water flow models to estimate the part of an unconfined aquifer that interacts with a lake. Taniguchi (2001) used a one-dimensional advection-dispersion model calibrated to chloride data to determine that b for Lake Biwa, Japan, was 150 meters. Lee and Swancar (1997) used vertical ground-water flow divides to determine b for their flow-net analysis for a lake in Florida. Perhaps the most thorough investigation to date is a study of flow between two lakes in northern Wisconsin. Flow-net, isotopic and geochemical, and numerical modeling approaches have been used to determine the relative volumes of water that flow from the upgradient lake to the downgradient lake and water that flows from the upgradient lake, beneath, and ultimately beyond the downgradient lake (for example, Kim and others, 1999).

Conceptual models of hypothetical settings can be useful in constraining estimates of exchange between ground water and surface water when sufficient field data are not available. Simply knowing the size, shape, and depth of a lake relative to its watershed can aid in determining the degree of interaction between the lake and its watershed. Two-dimensional and three-dimensional numerical and analytical tools can visually present the types and relative scales of flow paths associated with exchange between ground water and surface water (Townley and Davidson, 1988; Nield and others, 1994; Townley and Trefry, 2000). Recent updates of ground-water flow models allow more realistic simulation of exchanges between ground water and surface water than was previously possible. Hunt and others (2003) provide an overview of the usefulness of these improvements associated with the U.S. Geological Survey MODFLOW model (Leake, 1997; Harbaugh and others, 2000; Harbaugh, 2005).

Compared to errors associated with conceptualizing flow paths and determining aquifer properties, the remaining sources of error listed here usually are relatively minor. They are included, however, for completeness, and because in some situations they can represent a significant part of the total error associated with quantifying flow between ground water and surface water.

Measurement Error

Errors in making water-level measurements in wells and in observing surface-water stage generally are not significant relative to errors in determining K or A. Errors associated with determining the elevation of the top of the well casing relative to surface-water stage also typically are small. Increasing accuracy and availability of global positioning systems are reducing errors associated with determining well location. These errors can be significant, however, if the well is within a few meters of the surface-water body or if hydraulic gradients are very small. In this instance, greater care and more accurate methods should be used in determining the position of the well and the elevation of the top of the well casing relative to the surface-water stage.

Improperly Constructed Wells

Water-table wells in which water levels will be measured to calculate fluxes between ground water and surface water should be constructed so the water level in the well represents the phreatic surface of the aquifer (the water table). The screened interval of the well should be placed so it intersects the water table over the expected range of water-table fluctuations. Typical well-screen lengths for water-table monitoring wells range from 0.3 to 3 meters. Wells with long screens will integrate hydraulic head over the length of the well screen, and wells with short screens that are placed substantially below the water table will provide hydraulic head at depth in the aquifer that may be considerably different from the water-table head, especially within two to three aquifer thicknesses from the lake (Hunt and others, 2003).

Improper well construction also can alter hydrologic representation, particularly if the completion method results in the well screen being isolated from the aquifer. If drilling mud is used during well construction, for example, the well must be sufficiently developed following completion of the drilling to ensure that the well is in good hydraulic connection with the aquifer. For wells that are driven or pounded to the desired depth, a common installation method near the shoreline where the depth to water is shallow, care also needs to be given to proper development of the well. Well screens often are smeared with fine-grained sediment during the driving process and can be completely clogged if they are not flushed following installation. Hand-augered wells commonly are installed with the bottom of the well screen a short distance below the water table. It is difficult to auger through sand much beyond 1 meter below the water table because the sand collapses into the part of the hole below the water table. The consequence of the water table dropping below the bottom of the well screen is a dry well. A word of caution is in order for water-level measurements in wells constructed so the screen does not extend all the way to the well bottom (that is, when an impervious cap or drive point extends beyond the bottom of the screen): a small amount of water can be trapped inside the cap or drive point and remain in the well even if the water table has dropped below the bottom of the well. In such instances, the observer can still make a water-level measurement in the well and may not realize that the actual water table is below the bottom of the well.

The well screen also needs to be selected with a slot size (width of the openings in the screen) that is appropriate for the geologic material in which the screen is installed. If the slot size is too small, water levels in the well will lag behind changes in hydraulic head in the aquifer (Hvorslev, 1951). If the slot size is too large, particles will pass through the screen and may fill the well bore. Improper slot size may not be important when monitoring water levels on a weekly or less frequent interval, but can be very important if water-level change is recorded as part of a slug test or aquifer test.

Improperly Maintained Wells

Water-table monitoring wells can become clogged with sediments or bacterial growth, in part because so little water typically flows through a monitoring-well screen. Chemical precipitates (scale) also can clog the openings of a well screen. These processes decrease the connectivity of the well with the aquifer, creating a delayed response between the water level in the well and the hydraulic head in the aquifer. In extreme instances, the water level in the well becomes unresponsive to temporal changes in aquifer hydraulic head. Monitoring wells should be flushed occasionally to test and maintain connectivity with the aquifer.

The top of the well casing is vulnerable to accidental damage or vandalism. Protective devices for wells should be maintained and records kept in order to document any changes in the elevation of the top of the well casing to which water levels in the well typically are referenced.

Unstable Wells and Staff Gages

Water-table monitoring wells located near surface-water bodies commonly are quite shallow because the depth to the water table is shallow. Shallow well casings can move vertically in response to pumping for water-sample collection, frost, and settling of well cuttings placed in the annular space between the well casing and undisturbed sediments. This is particularly common for wells installed in wetland sediments. Shallow wells constructed with plastic casing can break from ice expansion during subfreezing temperatures. Wells and surface-water staff gages located near a downwind shoreline also can be tilted, moved horizontally, or broken if surface ice is pushed onto the shoreline during fall freeze or spring thaw. For longer term studies, it may be cost effective to install a sturdy surface-water monitoring station so that sources of environmental damage are minimized (Buchanan and Somers, 1982). A less expensive means for obtaining greater stability in a surface-water stage record is the installation of a siphon gage that allows measurement of surface-water stage in a protected environment (McCobb and others, 1999). Annual leveling surveys are necessary for surface-water staff gages, as well as many near-shore wells, in order to document changes in the elevation of the staff gage or the top of the well casing. Multiple survey benchmarks can aid in maintaining long-term elevational accuracy for staff gages and shallow, near-shore wells.

Violation of Underlying Assumptions

The previously discussed assumptions of homogeneity and isotropy, inherent in most calculations of exchange between ground water and surface water, are rarely met or appropriate for near-shore settings and can result in large errors in quantifying exchange between ground water and surface water. Assumptions of two-dimensional areal flow also typically are violated in near-shore regions (two to three aquifer thicknesses from the surface-water feature) where convergences and divergences of flow lines are common. The Darcy approach also assumes that the system is in a steady-state condition. Although the natural world is rarely if ever at true steady state, the system often will have periods when water levels are not changing appreciably over time, during which representative estimations of average flows can be made. It often is instructive to construct a simple computer model of the physical hydrologic setting, even if the entire hydrogeologic framework is not adequately known. Such a tool facilitates testing of the significance of one or more of these assumptions. In addition, a preliminary model can be used to help identify sensitive parameters and locate areas in the watershed where additional data collection would be most beneficial.

Hydraulic Potentiomanometer

The hydraulic potentiomanometer, sometimes referred to as a mini-piezometer, is a portable drive probe connected to a manometer (fig. 5). The manometer provides a comparison between the stage of a surface-water body and the hydraulic head beneath the surface-water body at the depth to which the screen at the end of the probe is driven (Winter and others, 1988). The difference in head divided by the distance between the screen and the sediment-water interface is a measurement of the vertical hydraulic-head gradient. By driving the probe to different depths beneath the sediment-water interface, the probe can provide information about variability in vertical hydraulic-head gradient with depth. The device does not give a direct indication of seepage flux, but when used in combination with a seepage meter, which does measure water flux, the two devices can yield information about the hydraulic conductivity of the sediments (for example, Kelly and Murdoch, 2003; Zamora, 2006). Because this device provides a quick characterization of the direction and magnitude of the vertical hydraulic gradient, it is useful as a reconnaissance tool in lakes, wetlands, and streams. It also is useful in areas where near-shore water-table wells or piezometers do not exist, are sparsely distributed, or are impractical to install and maintain.

The original hydraulic potentiomanometer design (fig. 6) consists of two nested stainless-steel pipes separated by O-rings that rest in grooves machined into the inner pipe. A screen with a machined point is threaded onto the inner pipe. The outer pipe acts as a shield for the screen; it covers

Figure 5. Hydraulic potentiomanometer showing drive probe inserted into lakebed and manometer indicating a very small vertical hydraulic-head gradient (blue arrows indicate water levels on manometer). (Photograph by Donald Rosenberry, U.S. Geological Survey.)

the screen and prevents damage to the screen and smearing of fine-grained sediments during insertion of the probe. A manometer is connected to the probe to allow measurement of the difference between head at the exposed well screen and the stage of the surface-water body.

Once the probe is pushed to a desired depth beneath the sediment-water interface, the outer pipe is retracted to expose the screen. At this point, one could simply measure from the top of the well pipe to the water level inside the probe and to the surface-water level outside of the probe. The difference between these measurements is the head difference. For convenience, and to better resolve small head differences, a manometer is attached to the probe. A vacuum is applied at the top of the manometer, pulling water through tubing connected to the probe and the surface water. Greater resistance of flow through the well screen may require that the surface-water tube be clamped to allow development of sufficient suction to pull water through the well screen and tubing. When all of the tubing is full of water and free of bubbles, air is bled into the top of the manometer until the menisci are visible in the tubing on both sides of the manometer (fig. 5). The difference in height of the menisci equals the difference between head at the screen in the sediment and the stage of the surface-water body.

The hydraulic potentiomanometer works well in fine sands and coarser materials. It becomes difficult to pull water through the screen if the sediments contain significant amounts of silt, clay, or organic deposits. The probe is difficult to insert in rocky or cobbly sediments because of the difficulty of driving the probe past the rocks and also because it is difficult to obtain a good seal between the outer pipe of the probe and the sediments. Rocks and cobbles near the shoreline often are only a surficial veneer; however, a measurement usually is possible if the probe can penetrate the surface layer.

Variability in the direction and magnitude of horizontal hydraulic-head gradient with distance from shore can be determined by making measurements along transects oriented perpendicular to the shoreline. The probe should be inserted to the same depth beneath the sediment-water interface at each measurement location. Otherwise, it is impossible to distinguish spatial variability in horizontal gradients from spatial variability in vertical gradients. One end of each transect typically extends to the shoreline, but measurements also can be made onshore in places where the probe can be driven deeply enough to reach the water table. For onshore measurements, the hydraulic potentiomanometer probe provides data equivalent to that of a shallow, near-shore monitoring well, while the tubing in the lake serves as a surface-water gage. Where the near-shore land-surface slope is small, the probe can be inserted a considerable distance from the shoreline, although the tubing needs to be long enough to extend from the well probe to the surface-water body. The vertical distribution of vertical hydraulic-head gradients can be determined by driving the probe to multiple depths beneath the sediment-water interface at each measurement location. This provides information about geologic heterogeneity with depth beneath the sediment-water interface, which can have a large influence on depth-integrated hydraulic-head gradients. In rivers, it is common for sediments to be composed of alternating layers of organic and inorganic sediments or fine-grained and coarse-grained sediments. Measurements often cannot be made in the organic or fine-grained layers, in which case measurements should be attempted in the more permeable layers. Differences in head between the transmissive layers often are large because the intervening low-permeability layers limit the equalization of pressure between the transmissive layers. Biogenic gas, which is common in many riverine sediments, can make obtaining bubble-free measurements difficult.

Differences in hydraulic head, although dependent on the depth to which the probe is inserted, typically range from 0 to 10 centimeters, but head differences as much as 30 centimeters are not uncommon. In some settings, the head difference can be very large, primarily because of local-scale geologic heterogeneity. In rare instances, head differences are greater than the length of the manometer, in which case the manometer can be raised, allowing the lower-head meniscus to be situated in the clear flexible tubing connected to the base of the manometer. For example, a head difference of approximately 2.4 meters was reported at a site where water was flowing from a lake to ground water (Rosenberry, 2000). The extreme gradient was present because a nearby lake had a water level 14 meters lower than the upper lake. Coarse sand was present between the two lakes, and much of the head difference between the two lakes was distributed across a 20-centimeter-thick layer

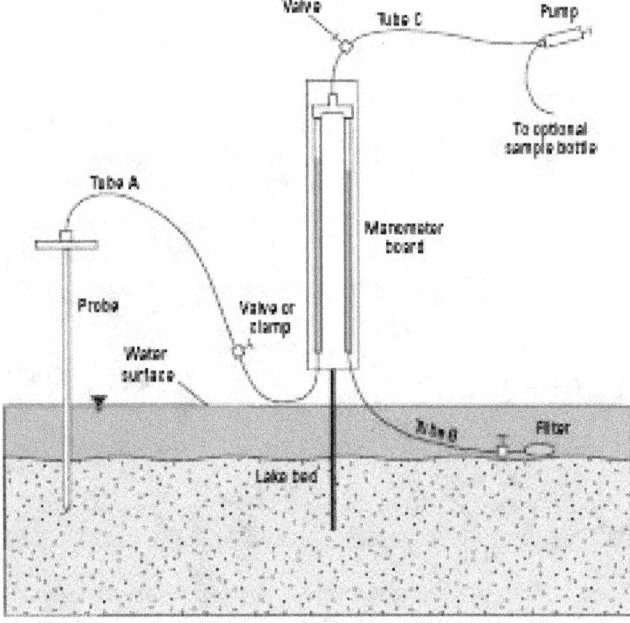

Figure 6. Components of the hydraulic potentiomanometer system. (Modified from Winter and others, 1988; copyright 1988 by the American Society of Limnology and Oceanography, Inc., used with permission.)

of organic-rich, sandy sediment at the sediment-water interface of the upper lake. Although still quite permeable compared to other lake sediments in the area, the permeability of the top 20 centimeters of sandy lake sediment was one to two orders of magnitude lower than that of the underlying coarse sand. The manometer had to be raised well above the lake surface in order to measure the large head difference.

Early versions of the hydraulic potentiomanometer used a vacuum bottle for collection of the water because hand-held pumps did not work well if they became wet (fig. 6). However, hand-cranked or motorized peristaltic pumps work well for pulling water through the manometer system. A cordless drill attached to a peristaltic pump head also can be used as a portable pump (J. Lundy, Minnesota Department of Health, oral commun., 2005). For a small-volume well and manometer system, a large syringe can serve as a pump (D.R. LeBlanc, U.S. Geological Survey, oral commun., 2005).

Water samples can be collected with the hydraulic potentiomanometer. Many investigators choose to bypass the manometer when collecting samples to minimize the potential for sample contamination from the tubing.

Sources of Error

Several sources of error attend the use of a hydraulic potentiomanometer:

1. Measurement error,
2. Improper leveling of the manometer,
3. Unstable hydraulic head,
4. Improper seal between outer pipe and the sediments,
5. Large bubbles entrained in tubing,
6. Leaks or clogging, and
7. Waves, standing waves, and seiches.

Each item listed above is discussed in detail below.

Measurement Error

Errors in measurement can result from improperly reading the menisci on the manometer. For very small differences in head, a common occurrence in highly permeable sediments, this error can result in a misinterpretation of the direction of flow of water across the sediment-water interface. Capillarity typically is not an issue unless very small diameter plastic tubing is used or the tubing diameters on the manometer are different. Hydrophobicity, however, may become significant if small-diameter plastic tubing is used, in which case the water menisci in the tubing may resist movement in response to small changes in hydraulic-head gradient.

Head differences can be amplified by use of a light oil in place of air at the top of the manometer (Kelly and Murdoch, 2003). The degree of amplification depends on the density of the light oil relative to the density of water:

$$dh = dh_{oil}\left(\frac{\rho_w - \rho_{oil}}{\rho_w}\right). \qquad (4)$$

where

dh is difference in hydraulic head over the distance between the sediment-water interface and the piezometer screen (L),

dh_{oil} is difference in elevation between the oil-water interface on the piezometer side of the manometer and oil-water interface on the surface-water side of the manometer (L),

ρ_w is density of water (M V^{-1}),

and

ρ_{oil} is density of oil (M V^{-1}).

Kelly and Murdoch (2003) used vegetable oil with a density of 0.9 gram per cubic centimeter, which increased the head difference tenfold.

Another source of measurement error is the determination of the depth to which the screened interval of the probe is inserted. An easy solution for determining this depth is to use an engraving tool to mark the outer pipe with depth increments. This distance should be recorded before the outer pipe is retracted to expose the screen. Distance typically is relative to the center of the screened interval of the probe.

Improper Leveling of the Manometer

This common problem becomes important when the two sides of the manometer are separated by a considerable distance (as is the case with the manometer shown in fig. 5), or when the difference in head is small. Out-of-level error can be minimized by installing a bubble level on the manometer and by constructing the manometer so the two parallel tubes are positioned close to each other (fig. 7).

Unstable Hydraulic Head

Most measurements of difference in head stabilize in a matter of seconds to minutes. In low-permeability sediments, it can take from tens of minutes to hours for head at the probe screen to stabilize. In such cases, observations of difference in head are repeated until the difference in head stops changing, indicating stabilization. Stabilization time also can provide a relative indication of the permeability of the sediments at the location of the probe screen.

Improper Seal Between Outer Pipe and the Sediments

If the hydraulic potentiomanometer is inserted in rocky or gravelly sediments, or if the probe is not inserted cleanly into the sediments (that is, if the probe is rocked back and forth during insertion), or if the probe is inserted a very short distance into the sediments, water can flow vertically along the

Water level

Water level

Figure 7. Hydraulic potentiomanometer designed to place the manometer tubes connected to the drive probe and to the surface-water body close together to minimize out-of-level errors. (Photograph by Jim Lundy, Minnesota Department of Health.)

Larger scale areas of unsaturated sediments also have been identified, in which case the tubing from the probe is filled primarily with air and very little water. Although it is not possible to measure a hydraulic-head difference in these instances, the hydraulic potentiomanometer remains useful in that it can identify these sometimes unexpected hydrologic conditions. Rosenberry (2000) reported a large, apparently permanent wedge of unsaturated sediments beneath the edge of a lake that was identified on the basis of measurements made with the hydraulic potentiomanometer. This unsaturated sediment was in direct connection with the adjacent unsaturated zone onshore and extended up to 20 meters beyond the shoreline of the lake. The hydraulic potentiomanometer also was used at a small pond to determine the vertical and areal extent of pockets of gas beneath the pond that were several meters in diameter. The gas likely was trapped when the pond stage rose and the shoreline rapidly moved laterally to cover formerly unsaturated near-shore sediments (Rosenberry, 2000).

Leaks or Clogging

Leaks can form (1) at the O-rings that separate the inner and outer pipes, (2) between the inner rod and the tubing to which it is connected, and (3) between the tubing and the manometer. Leaks also can occur within the manometer plumbing. Leaks can cause formation of bubbles in the water contained within the probe and tubing, which can cause the manometer to indicate an erroneous difference in head. O-ring leaks can be prevented by liberal use of O-ring grease. Other leaks can be eliminated by using clamps, sealant or tape. The entire system can be clogged if the well screen is torn or absent and sediments are pulled through the tubing. Clogging also is likely if the end of the surface-water tubing settles into the bed sediments. A screen can be placed over the end of the surface-water tube to prevent sediments from entering the tubing. Also, a weight often is applied to the surface-water tube to keep the tube from floating to the surface and allowing air to be pulled through the tubing.

outer surface of the probe, with the result that the difference in head between the screen and the surface water is less than the actual difference. This "short circuiting" of head can be prevented by driving the probe straight into the sediments, or by driving the probe farther into the sediments.

Large Bubbles Entrained in Tubing

At many sites, biogenic gas is pulled through the probe and is visible in the tubing connecting the probe to the manometer. Gas bubbles inside the tubing can change volume with a change in temperature and thereby corrupt the difference in head displayed by the manometer. Care should be taken to ensure that large bubbles (large enough to extend across the entire cross section of the tubing) are removed prior to bleeding air back into the top of the manometer to take a reading. Very small bubbles also may appear when a strong vacuum is applied to pull water through the well screen. These bubbles are the result of the water degassing in response to the suction pressure. Typically, they do not present a problem because they occupy a very small volume, but over time they may grow as the water warms. The problem can become significant with increased equilibration time.

Occasionally, small lenses or zones of sediments beneath surface-water bodies are unsaturated, commonly because of discrete pockets of gas generated from organic decomposition.

Waves, Standing Waves, and Seiches

Difference in head between the surface-water body and the screened interval of the hydraulic potentiomanometer can vary with short-term changes in surface-water stage caused by waves, seiches, or even standing waves in fast-moving streams or rivers. Waves make it difficult to make a measurement if the head difference is small. The surface-water tube can be placed inside a small stilling well (even something as simple as a coffee can) with holes drilled in the side to dampen stage fluctuations from waves. Seiches (internal waves) are common on large lakes and rivers and can be dealt with by making measurements at the same location multiple times over a period that is appropriate for the periodicity of the seiche.

Other Similar Devices

Numerous other devices have been constructed to measure difference in head between surface-water bodies and the underlying ground water, involving a modification of the probe, the method for measuring head difference, or both. Squillace and others (1993) modified the hydraulic potentiomanometer by making the probe longer and adding a drive hammer in order to place the screened interval at depths as great as 3 meters below the sediment-water interface. This device was used by Rosenberry (2000) to determine the horizontal and vertical extent of unsaturated sediments beneath a lake (fig. 8). Another drive-hammer device has the manometer connected to the drive probe to minimize components that need to be carried in the field (fig. 9). Mitchell and others (1988) clamped the well and lake tubing to a metric ruler to create a simple manometer for making measurements of difference in head.

Members of the Cullen Lakes Association in northern Minnesota modified a well probe to eliminate the manometer. They used a "mini dipper" small-diameter electric tape to make measurements of depth to water inside and outside the probe. The measurement outside of the probe was made through a length of semirigid tubing; the top of the tubing was flush with the top of the probe, and the bottom extended to the surface water (fig. 10) (Ted P. Soteroplos and William (Bill) J. Maucker, Cullen Lakes Association, written commun., 1995). A commercially available, retractable, stainless-steel soil-gas vapor probe was used to avoid having to manufacture a retractable well screen.

Several other small-diameter devices also have been developed to measure vertical-head gradients beneath surface-water features. Lee and Cherry (1978) describe the use of a flexible plastic tube with a screen attached to the end. The tube is driven to depth inside a larger diameter rigid steel pipe that is removed once the insertion depth is reached, allowing the sediment to collapse around and seal the tube in place in the sediment. With the tubing extended above the surface, the water level inside the tube is compared to the surface-water stage. More recently, a root-watering device, commonly available at hardware stores, has been used to measure vertical hydraulic-head gradients beneath surface-water bodies (Wanty and Winter, 2000). A coil of tubing is connected to the top of the probe, and when positioned properly with respect to the water surface, is used to indicate difference in head between that in the probe and that of the surface-water body (fig. 11). A commercially available probe (MHE PP27) is used to collect water samples and to make measurements of difference in head beneath the sediment-water interface in much the same method (Henry, 2000). This device also makes use of clear tubing placed at the water surface to measure difference in head (fig. 12).

Several investigators have developed methods for determining head gradients at multiple depths beneath the sediment-water interface. Duff and others (1998) designed a device for collecting water samples from multiple depths beneath a streambed. If clear tubing is used, hydraulic heads also can be related to stream stage. Lundy and Ferrey (2004) used a combination of drive points and multilevel samplers that could be left in place for the duration of the study. Their study design allowed repeat measurement of head gradients and collection of water samples so the investigators could determine the extent and growth of a contaminant plume that intersected a stream. Both devices allowed rapid measurements and convenient collection of water samples from multiple depths.

Figure 8. Hydraulic potentiomanometer probe with drive hammer shown driven about 2 meters beneath the lakebed. Manometer and hand-crank peristaltic pump are visible in background. (Photograph by Donald Rosenberry, U.S. Geological Survey.)

Figure 9.   Hydraulic potentiomanometer (created by Joe Magner, Minnesota Pollution Control Agency) with manometer connected to drive probe. Note the proximity of the lake and drive-point tubes to minimize out-of-level errors. Note also the in-line water bottle to keep the vacuum pump dry. (Photograph by Donald Rosenberry, U.S. Geological Survey.)

If a considerable amount of trapped gas is encountered, thus making it difficult to get a bubble-free measurement, it is sometimes possible to pull water rapidly through the screen, evacuating much of the gas from the sediments near the probe screen. After waiting a few minutes, water then can be pulled slowly through the screen without pulling additional gas bubbles into the tubing.

The screened interval commonly will break when using a drive hammer to position a hydraulic potentiomanometer probe, especially if many blows are required and the probe is made from stainless steel. Stainless steel is relatively brittle, and the many holes drilled in the screened interval weaken the metal tube, which may lead to failure from the shock of the drive hammer. It is advisable to build the device with the screen as a separate part that is threaded onto the interior rod of the probe, so damaged or broken screens can be removed and replaced. It also is advisable to tighten the screen frequently because the shock of driving the probe often loosens the threads connecting the screen to the rest of the probe.

The screen should be retracted inside the outer sheath before removing the probe after a measurement has been completed. This prevents the screen from being damaged during removal of the probe and also traps the sediment that surrounds the screen while making the measurement. This allows a qualitative description of the sediments at the depth at which the measurement is made.

Calm surface-water conditions are required for all of these designs that make use of a length of clear tubing inserted into the surface-water body. A manometer could be used with any of these devices, although most of these alternative approaches were developed to avoid use of a manometer in order to simplify the measurement system.

Cautions and Suggestions Related to Use of the Hydraulic Potentiomanometer

Buried debris, such as logs, rocks, and even old tires, often is encountered when driving the hydraulic potentiomanometer probe. In most instances, it is possible to reinsert the probe 0.5 meter away and drive the probe to the desired depth.

The observer should not stand within 1 meter of the probe when making a measurement. The weight of the observer can compact sediments and cause a several-centimeter change in the measured head difference. This artifact is especially notable in soft sediments.

Seepage Meters

The seepage meter is one of the most commonly used devices for making a direct measurement of the flux of water across the sediment-water interface. Early versions were developed to measure water losses from irrigation canals (Israelson and Reeve, 1944; Warnick, 1951; Robinson and Rohwer, 1952; Rasmussen and Lauritzen, 1953). Many of these devices were expensive and unwieldy and were little used beyond the application to canals. Carr and Winter (1980) provide an annotated bibliography of the early literature on seepage meters, including drawings of some of the devices. Lee (1977) developed an inexpensive and simple meter that has changed little during the decades since its inception. Lee's meter consists of the cut-off end of a 208-liter (55-gallon) storage drum, to which is attached a plastic bag that is partially filled with a known volume of water (fig. 13). The drum, or chamber, is submerged in the surface-water body and placed

Figure 10.   Portable well probe consisting of a commercially available retractable soil-gas vapor probe connected to threaded pipe with tubing inside the pipe connected to the vapor probe. A separate tube taped to the outside of the pipe extends to the lake-water surface. (Photograph by Donald Rosenberry, U.S. Geological Survey.)

that values less than 0.01 centimeter per day (Lee and Cherry, 1978), 0.04 centimeter per day (Harvey and others, 2004), or 0.08 centimeter per day (Cable and others, 1997a) are too small to be measured accurately. A recently developed meter designed for use in benthic ocean settings is capable of measuring exceptionally slow seepage rates as small as 3×10^{-5} centimeters per day (Tryon and others, 2001). Several values of 100 centimeters per day or greater have been reported (100 centimeters per day—Asbury, 1990; 130 centimeters per day—Belanger and Walker, 1990; 240 centimeters per day—Rosenberry, 2000; 275 centimeters per day—Paulsen and others, 2001). Duff and others (1999) measured a flux of nearly 5,200 centimeters per day from a 2- to 3-centimeter-diameter, boiling-sand spring in a small stream in northern Minnesota.

The half-barrel seepage meter is relatively easy to use and conceptually simple to operate. The cylindrical seepage chamber (with bag detached) first is placed on the submerged sediment and slowly inserted into the bed with a twisting, sediment-cutting action. Care must be taken to ensure a good seal between the chamber and the sediment. Buried rocks

in the sediment to contain the seepage that crosses that part of the sediment-water interface. The bag then is attached to the chamber for a measured amount of time, after which the bag is removed and the volume of water contained in the bag is remeasured. The change in volume during the time the bag was attached to the chamber is the volumetric rate of flow through the part of the bed covered by the chamber (volume/time). The volumetric rate of flow then can be divided by the approximately 0.25-square-meter area covered by the chamber to express seepage as a flux velocity (distance/time). Flux velocity is useful because it normalizes the area covered by the seepage meter and allows comparisons of results with other studies (and other sizes of seepage meters). Seepage flux velocity typically is multiplied by a coefficient that compensates for inefficiencies in flow within the meter, restrictions to flow through the connector between the bag and the chamber, and any resistance to movement of the bag as it fills or empties.

The range of seepage rates that have been reported from coastal and fresh-water settings is approximately five orders of magnitude. Values as small as 0.01 centimeter per day have been reported (for example, Cherkauer and McBride, 1988; Yelverton and Hackney, 1986), although some studies indicate

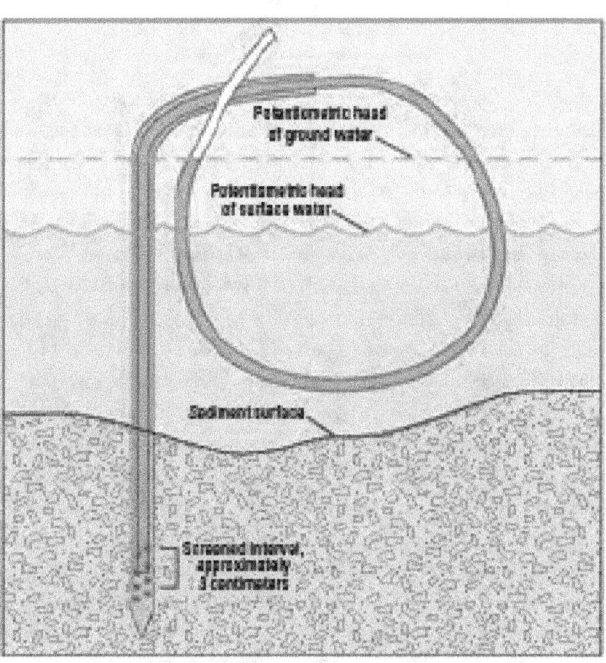

Figure 11.   Well probe constructed from a commercially available root feeder with the coiled tubing substituting for a manometer. (Modified from Wanty and Winter, 2000.)

PP27 used as a piezometer/manometer to determine the elevation of ground water relative to surface water

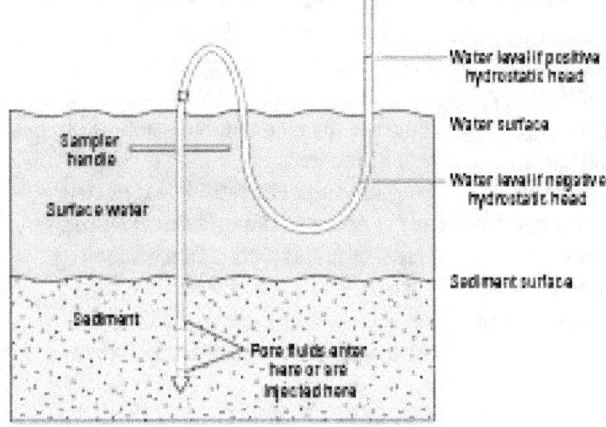

Figure 12.   MHE PP27 probe used to indicate difference in head (modified from Henry, 2000). (Photograph by Mark Henry, Michigan Department of Environmental Quality.)

Lee (1977) originally used the cut-off end of a 208-liter (55-gallon) drum, but many other types and sizes of chambers also have been used, including coffee cans (Asbury, 1990), inverted plastic trash cans (S.E. Hagerthey and D.O. Rosenberry, U.S. Geological Survey, written commun., 1998), lids from desiccation chambers (Duff and others, 1999), fiberglass domes cemented to a limestone bed (Shinn and others, 2002), and even galvanized stock tanks (Landon and others, 2001; Rosenberry and Morin, 2004). The size of the chamber should be selected for convenience and for the expected rate of seepage across the sediment-water interface. A large-diameter meter can measure more accurately an extremely small flow across the sediment-water interface, and it also better integrates small-scale spatial variability in seepage flux. A large meter, however, can be unwieldy and it also is more difficult to ensure a good seal in uneven, rocky, or debris-laden settings. Alternately, flow from several normal-sized chambers can be routed to one seepage bag to increase the surface area and integrate spatial heterogeneity (Rosenberry, 2005). Large-diameter seepage meters often are difficult to remove from the sediments following their use, as are smaller-sized chambers inserted into silty or clayey sediment. A simple solution is to insert a length of tubing inside of the chamber and blow air into the chamber until the buoyancy force lifts the chamber out of the sediments. Some users have installed additional openings in the top of the chamber that are opened prior to removal of the chamber in order for water to flow into the chamber as it is pulled from the sediments. Additional openings also reduce the chance for "blowouts" or sediment compression during chamber installation.

Much has been written regarding the type and size of the bag attached to the chamber (for example, Erickson, 1981; Shaw and Prepas, 1989; Cable and others, 1997a; Isiorho and Meyer, 1999). Bags as small as condoms (Fellows and Brezonik, 1980; Duff and others, 1999; Isiorho and Meyer, 1999; Schincariol and McNeil, 2002) to as large as 15-liter trash bags (Erickson, 1981) have been used, with 4-liter sandwich bags among the most common choices. Most plastic bags have a "memory effect" caused by the manufacturing process that results in a slight pressure created by the bag as it moves to a more relaxed position. This can result in errors in measurement that become substantial in low-flux settings. Shaw and Prepas (1989) reported an anomalous influx of water during the first 30 minutes following bag installation that they were able to eliminate by prefilling the bags with 1,000 milliliters of water. Cable and others (1997a) reported similar results. Shaw and Prepas (1989) suggested using a 4-liter-sized bag and adding a known volume of water (1 liter or more), even in settings where flow of water was from ground water to surface water, because these procedures tended to minimize the memory effect. Shaw and Prepas also suggested prewetting the bags prior to installation on the chamber to

or woody debris can prevent the edges of the chamber from extending into the sediment and may allow short-circuiting of water beneath the edge of the chamber. Some investigators have packed sediments around the outside of the chamber to create a good seal (Cable and others, 1997a). Occasionally, in sandy or gravelly settings, it is necessary to stand on the chamber and gently rock it back and forth to force it into the sediment. Sometimes this action is necessary in weedy settings where the meter needs to cut through a part of the weed bed in order to achieve a good seal. Harvey and others (2000) made circular vertical slits in the fibrous peat in order to install seepage meters in wetlands in the Florida Everglades. Standing on the chamber should be a last resort, however, because rapid emplacement can cause "blowouts" of the sediment adjacent to the chamber (Lee, 1977), or compress sediments beneath the chamber, and disturb the natural rate of water flow through the sediments. The chamber should be emplaced with a slight tilt so that the opening to which the bag is attached is near the uppermost edge of the meter, which facilitates the release of gas from the sediment. Sediments often are compressed beneath the seepage-meter chamber during meter insertion, and flow is temporarily disrupted. The sediments and flow need to equilibrate before a bag is attached. Substantial error can result if measurements are made too soon following meter installation.

A

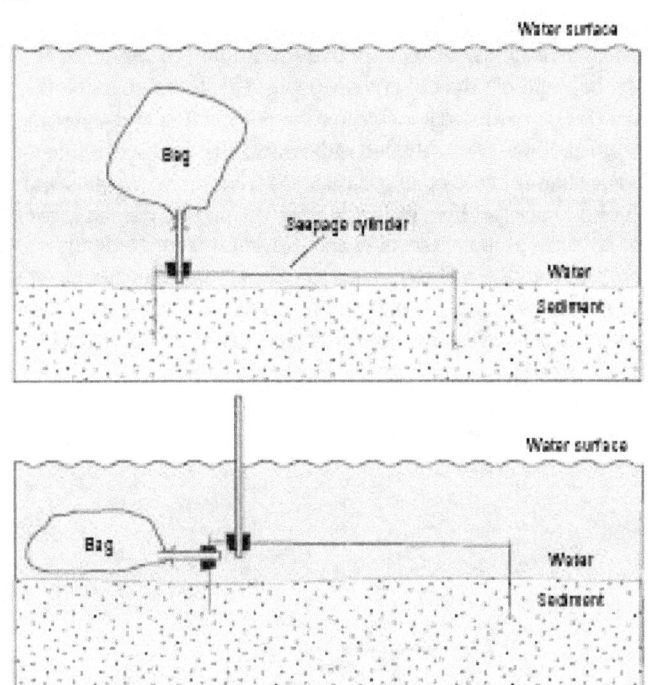

B

C

Figure 13.   *A,* Half-barrel seepage meter (modified from Lee and Cherry, 1978, used by permission of the Journal of Geoscience Education). The top panel shows typical installation with bag connected to a tube inserted through a rubber stopper. The bottom panel shows installation in shallow water with vent tube to allow trapped gas to escape. *B,* Standard half-barrel seepage meter in place in the field. (Photograph by Donald Rosenberry, U.S. Geological Survey.) *C,* Electromagnetic seepage meter (foreground) installed next to a half-barrel seepage meter. Cable extending from seepage cylinder connects to signal conditioner and power supply located on nearby anchored raft. (Photograph by Donald Rosenberry, U.S. Geological Survey.)

avoid bias related to the loss of water from adhesion of water to the inside of the bag. Blanchfield and Ridgway (1996) indicated that seepage rates were inflated by as much as one order of magnitude if unfilled bags were used instead of bags prefilled with 1,000 milliliters of water. Asbury (1990), reporting results from seepage measurements made where water was rapidly flowing from a lake to ground water, indicated that the sides of the bag came into contact when the volume of water was 500 milliliters or less, which caused a reduction of flow out of the bag. Murdoch and Kelly (2003) indicated that the hydraulic head necessary to fill a 3,500-milliliter seepage bag was smallest when the bag was initially empty, increased to a relatively constant value once the bag contained about 100 to 200 milliliters of water, and then increased rapidly when the bag was within 500 to 800 milliliters of being full. They also determined that the resistance to filling the bag depended on the bag thickness.

Several studies have reported a preference for thin-walled plastic bags to minimize resistance to flow to or from the bag. Others have reported problems with fish chewing holes in the bags and switched to thicker walled bags (Erickson, 1981). One solution to the fish problem is to place one bag inside another. If this is done, however, it is important to place small holes in the corners of the outside bag to allow water between the bags to drain prior to measurement and to allow air to escape from between the bags prior to bag insertion. Another solution is to place the bag in a shelter, which also serves the purpose of minimizing the effects of waves and currents, described later. Thick-walled bags also have been used; intravenous-drip bags or urine-collection bags are especially convenient because the tubing that extends from the bag already is attached. Recent studies, however, which are discussed in the following section on sources of error, indicate that bag resistance induces substantial error to seepage measurements, so the use of thick-walled bags should be avoided.

In addition, cautions have been issued regarding collecting water-quality samples from a seepage meter (Brock and others, 1982; Belanger and Mikutel, 1985). Because the residence time of water contained inside the seepage chamber or bag may allow the chemistry of the water to change, samples may not be representative of the chemistry of water discharging across the sediment-water interface.

Seepage meters have been modified for use in extreme environments. Cherkauer and McBride (1988) created a seepage meter that included a concrete collar for measurement of seepage in Lake Michigan, where energy from large waves would dislodge unmodified devices (fig. 14). Dorrance (1989) and Boyle (1994) each designed seepage meters for use in deep water. Both designs consisted of a seepage chamber connected via tubing to a seepage bag installed inside a separate

housing. Dorrance allowed the bag shelter to float on the surface, whereas Boyle suspended the bag housing a short distance beneath the surface. Both designs allowed servicing of the bag without the aid of a diver (fig. 15). Hedblom and others (2003) modified a meter to measure gas flux and water flux from shallow, contaminated sediments. The device contained a mylar bag for collecting gas released from the sediments, and it also contained long rods that were driven into the sediment to hold the seepage chamber and prevent it from gradually sinking into the soft sediments. Shallow-water seepage meters also have been used to measure flows near the shoreline where seepage rates often are large (Lee and Cherry, 1978) (fig. 13, lower panel). Lee and Cherry simply attached the bag to the side of the seepage chamber rather than to the top. A bag also could be attached to the side of a seepage cylinder that extends

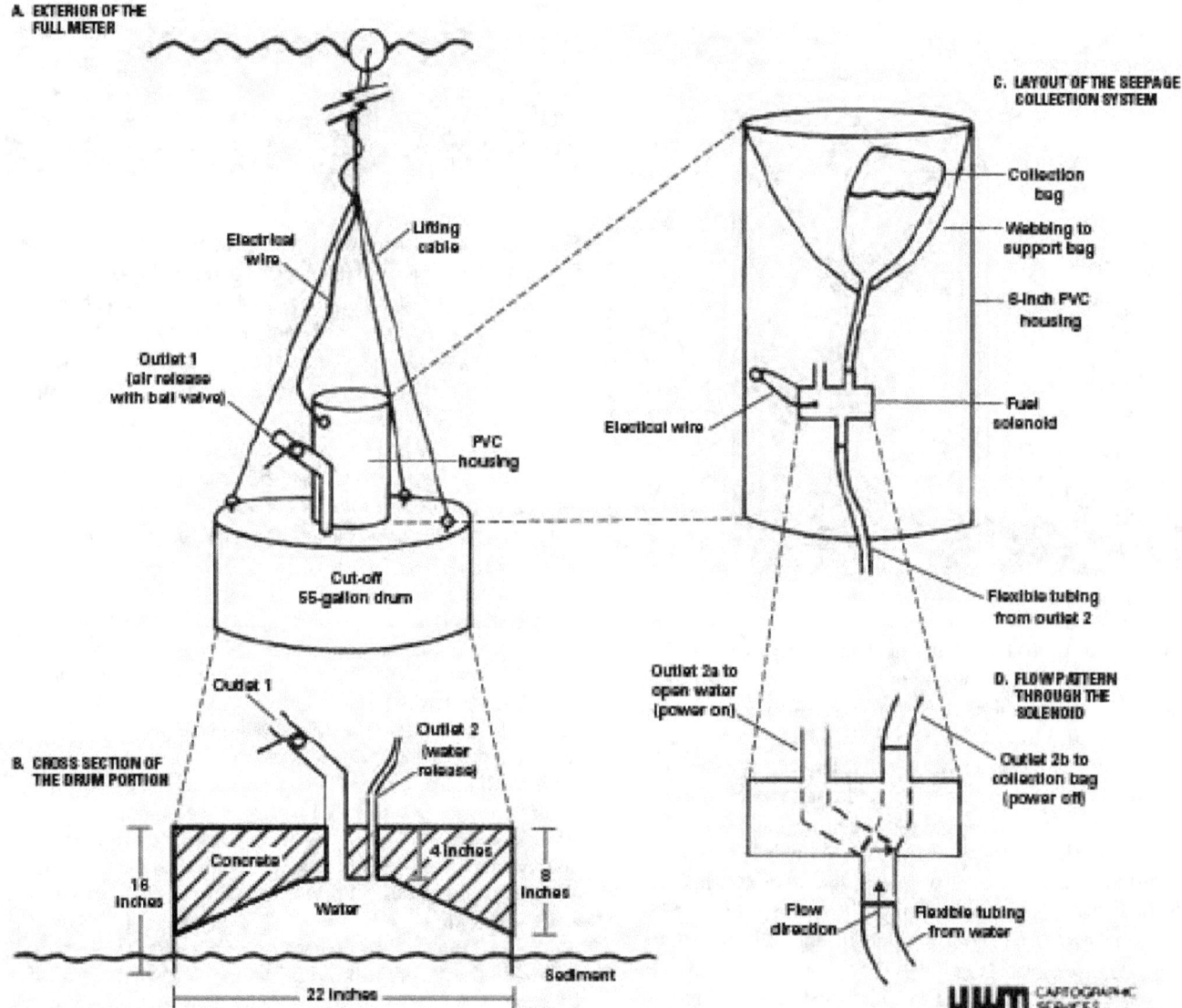

Figure 14. Seepage meter modified for use in large lakes (from Cherkauer and McBride, 1988. Reprinted from Ground Water with permission from the National Ground Water Association, copyright 1988).

above the water surface and contains a free water surface. Water then would flow into or out of the bag in order to maintain the same water level inside and outside of the seepage chamber. This procedure could allow measurement of seepage in the very shallow water closest to the shoreline where seepage rates often are the largest. Waves, however, could create potentially large flows through the submerged opening in the side of the chamber, leading to large measurement errors. Rosenberry and Morin (2004) reported instantaneous flow rates into and out of a near-shore seepage cylinder of more than 300 milliliters per second.

Seepage meters also have been modified for use in flowing water. In many streams and rivers, currents are sufficient to cause the bag to be deflected in a downstream direction, which may fold the bag across its connection to the chamber and either reduce or pinch off flow to or from the bag. Currents also may generate a pressure gradient across the bag membrane that could lead to erroneous measurements (described more completely in the next section on sources of error). Bags have been installed inside shelters to protect the bag from these currents. Designs have included shelters mounted to the top of the seepage cylinder (for example, Schneider and others, 2005) or shelters that rest on the sediment bed and are attached to the seepage cylinder via a short length of tubing or hose (for example, Landon and others, 2001). Bag shelters also protect the bag from wave action that may cause erroneously large seepage rates (Sebestyen and Schneider, 2001), and they maintain the bag in the proper orientation so it cannot swing with the current, which could pinch off the opening at the bag-connection point. David Lee (Atomic Energy of

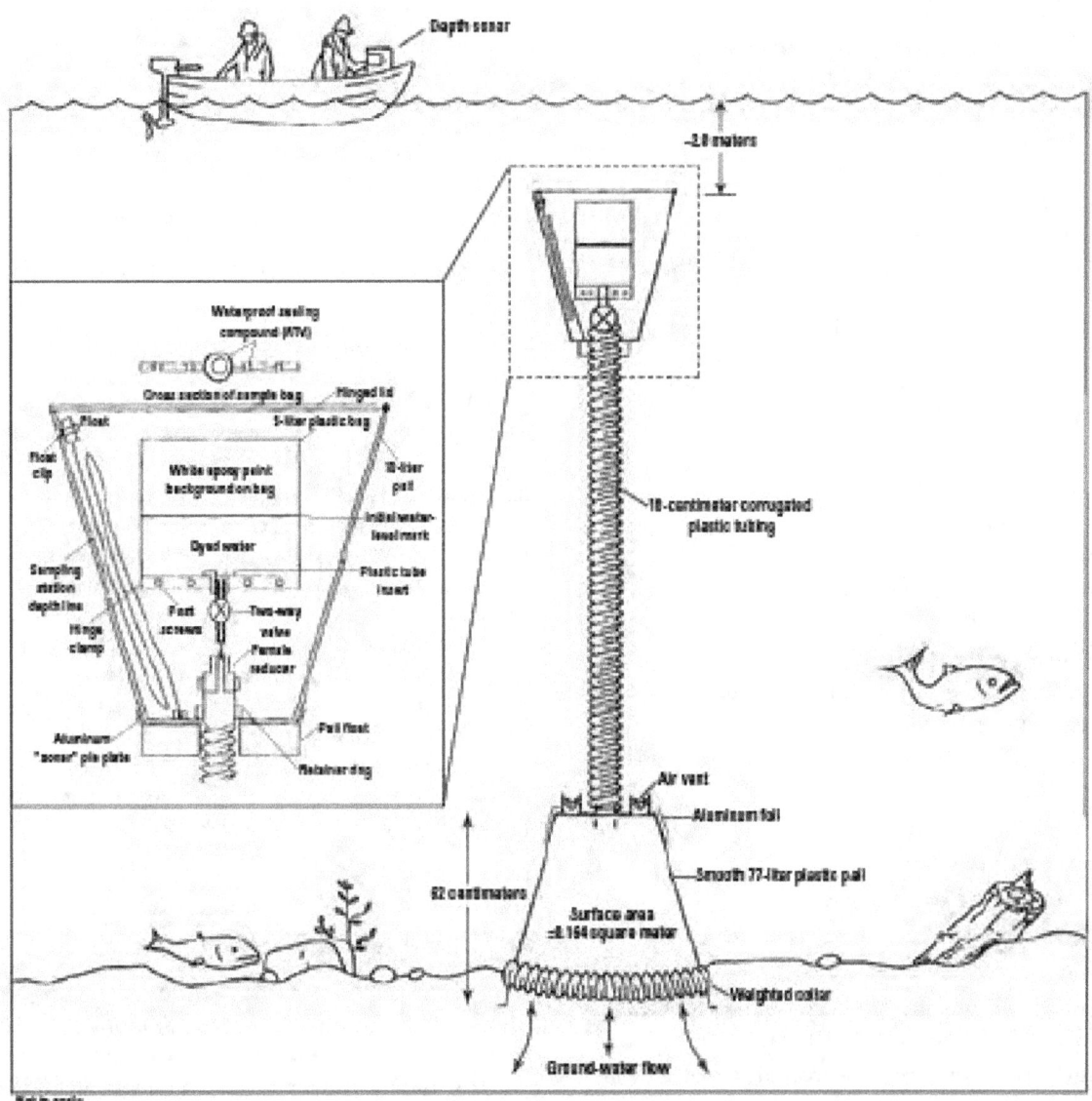

Figure 15. Ground-water seepage meter modified for use in deep water. (Modified from Boyle, 1994. Copyright 1994 by the American Society of Limnology and Oceanography, Inc., used with permission.)

Canada Limited, oral commun., 2006) suggested drilling small holes in the part of the tubing that extends inside of the bag to further prevent errors should the bag move to pinch off the end of the tube.

If a seepage meter is used in combination with the hydraulic potentiomanometer, local-scale values for vertical hydraulic conductivity can be obtained. A modification of particular interest is the "piezoseep" by Kelly and Murdoch (2003) that combines a seepage-meter chamber with a piezometer inserted through the center of the chamber. The authors replaced the seepage bag with a pump that pulls water at known rates from the area covered by the chamber. By accurately measuring the head difference between the piezometer point and the surface water in response to various pumping rates, they were able to calculate in-situ hydraulic conductivity. The relation between head gradient and seepage flux was determined for each meter, which allowed seepage rates to be monitored by simply measuring head differences at a desired time interval (Murdoch and Kelly, 2003).

Although numerous sources of error exist, especially associated with the seepage bag, the seepage meter is an attractive choice for quantifying flow across the sediment-water interface because of its simplicity and low cost. Perhaps as a result, care and training in the use and operation of seepage meters commensurate with their cost may have led to collection of poor-quality data for some studies. With proper understanding of the operating principles, knowledge of sources of error, and care in measurement, accurate determinations of flux across the sediment-water interface are possible, as has been demonstrated in many of the papers cited herein.

Sources of Error

Sources of error when using seepage meters include:

1. Incomplete seal between seepage-meter chamber and sediments, unstable cylinders;
2. Insufficient time between meter installation and first measurement;
3. Improper bag-attachment procedures, bag resistance, and moving water;
4. Leaks;
5. Measurement error;
6. Flexible seepage-meter chamber;
7. Insufficient or excessive bag-attachment time;
8. Accumulation of trapped gas;
9. Incorrect coefficient to relate measured flux to actual flux across the sediment-water interface; and
10. Insufficient characterization of spatial heterogeneity in seepage through sediments.

Each item listed above is discussed in detail below.

Incomplete Seal, Unstable Cylinder

Care should be taken to ensure that an effective seal exists between the seepage chamber and the sediments. After pushing the chamber into the sediments, one can feel around the base of the chamber to ensure that the bottom edge of the chamber cannot be felt. If the bed is rocky and a good seal is impossible, it may be possible to place a mud or bentonite seal against the edge of the meter to create a temporary seal. This practice, however, may introduce additional errors because the seepage immediately beyond the edge of the meter will be altered.

Meters installed in soft sediments also may be subject to a sealing problem of a different type. If sediments are not sufficiently competent to support the weight of the meter, the seepage chamber may slowly sink into the sediment following emplacement. This will displace water from inside the chamber that will flow into a seepage bag connected to the chamber. A solution to this problem is to use taller seepage chambers set deeper into the sediments (Fellows and Brezonik, 1980). If, for example, a 208-liter storage drum is used, it can be cut in half to make a seepage chamber with sidewalls that are about 45 centimeters tall. Another solution is to anchor the meter to rods driven deep into the sediments (Hedblom and others, 2003). Menheer (2004) designed a chamber with fins that rest on the bed surface so that the chamber is installed at a consistent depth in the sediments for every placement.

Insufficient Equilibration Time

This may be the most common source of error for scientists inexperienced in the use of seepage meters. It is tempting to install the seepage chamber and immediately begin making measurements. Sediments first need to be allowed to equilibrate following their compression during insertion of the seepage chamber. Time between chamber installation and first measurement typically is 1 day or more, but a few studies have reported waiting shorter times when working in sandy or gravelly sediments (table 2). Some investigations indicate that an equilibration time as little as 10 to 15 minutes is adequate (Lock and John, 1978; Lewis, 1987; Landon and others, 2001). Rosenberry and Morin (2004) used an automated seepage meter to demonstrate that most of the recovery to predisturbance seepage rates was achieved within 30 minutes after installation of the seepage chamber in a sandy lakebed.

Improper Bag-Attachment Procedures, Bag Resistance, and Moving Water

The procedure for attaching the bag to the seepage chamber depends on the attachment mechanism. With early designs, the bag was attached to a small-diameter tube that extended through a rubber stopper inserted into the chamber. It is important to not apply any pressure to the bag while pushing the rubber stopper into the chamber during bag

Table 2. Duration between emplacement of seepage meter and first installation of seepage-meter bag (equilibration time) from selected studies.

Reference	Site	Equilibration time
Lock and John, 1978	Lake Taupo, New Zealand	5–10 minutes
Landon and others, 2001	Platte River, Nebraska	10–15 minutes
Lewis, 1987	Coral reefs on Barbados	15 minutes
Rosenberry and Morin, 2004	Mirror Lake, New Hampshire	30–60 minutes
Libelo and MacIntyre, 1994	York River, Virginia	1 hour
Rosenberry, 2000	Lake Belle Taine, Minnesota	1–3 hours
Cable and others, 1997a	Gulf Coast, Turkey Point, Florida	At least 24 hours
Belanger and Kirkner, 1994	Mountain Lake, Florida	1 day
Erickson, 1981	Williams Lake, Minnesota	2 days
Shaw and Prepas, 1989	Narrow Lake, Alberta	2–3 days
Lee, 1977	Lake Sallie, Minnesota	Several days
Belanger and Montgomery, 1992	Laboratory tank tests	Several days
Shaw and Prepas, 1990a	Narrow Lake, Alberta	2–5 days
Boyle, 1994	Alexander Lake, Ontario	A few weeks

attachment. Significant volumes of water can be forced into or, more commonly, out of the bag during attachment. The same caution applies to removal of the bag from the chamber. A recent improvement in bag-connection design involves using a shutoff valve (Cable and others, 1997a) (fig. 16). A bag is attached to the shutoff valve and a fitting that connects to the threads of the shutoff valve is installed in the seepage chamber. Once the bag is properly filled and emptied of air, the valve can be closed for transport until the bag is threaded onto the meter, at which time the valve is opened and the measurement period begins. Upon measurement completion, the valve is closed and the bag removed for final volume measurement. This minimizes the possibility of the investigator inadvertently causing flow into or out of the bag during insertion and removal. Other connectors that do not require threads also can be used, but the user should test the connector to make sure that it does not leak under near-zero pressure conditions.

As mentioned previously, use of thin-walled bags has been recommended in numerous seepage studies. Thick-walled bags generate a greater resistance to inflation or deflation, and a larger head gradient is required to effect a change in volume inside the bag. Because some bags are constructed with a tube already in place (for example, intravenous bags, urine-collection bags, solar-shower bags), several studies have reported use of these bags in seepage-meter studies. Unless a calibration coefficient is determined for measurements made with these bags, however, it is likely that the measured seepage rates will substantially underestimate fluxes across the sediment-water interface. Murdoch and Kelly (2003) reported that thicker bags required a much larger correction multiplier (1.88) compared to thin-walled bags (1.25); they also reported the measurement variance for thick-walled bags was greater than for thin-walled bags. Rosenberry and Menheer (2006) reported similar values, ranging from a correction multiplier of 0.95 for thin-walled bags to 1.89 for a solar-shower bag. Based on these observations, use of thick-walled bags is not recommended. Use of condoms as seepage-meter bags also is not

recommended because they present a temporally variable bag resistance that is particularly difficult to account for during calibration for (Schincariol and McNeil, 2002).

Bags should be free of air bubbles prior to bag attachment. Bubbles exert a buoyant force on the bag, which can place a strain on the bag and either cause an artificial gain or loss of water in the bag as it deforms in response to the buoyant force, or prevent the bag from readily inflating or deflating to accommodate seepage gains or losses. Harvey and others (2000) indicated that excessive gas collected in a seepage bag led to artificially large fluxes of water into the bag. They designed their seepage meters so that the top of the bag rested on the water surface, which eliminated buoyant forces. A simple way to remove air from inside the bag is to pull the bag beneath the surface of the water body (or beneath the water surface in a bucket), with the opening of the bag pointing away from the water surface. As the bag is pulled beneath the water, air escapes through the opening. The bag can be pulled almost completely beneath the water surface until the opening of the bag is about to be submerged, at which point the opening is closed and the inside of the bag is virtually free of air.

Errors also can be introduced when the observer wades out to the seepage meter to attach the bag. In soft sediments, the weight of the observer standing next to the seepage meter may cause displacement of water from the sediments. This can be a problem even in sandy sediments. Rosenberry and Morin (2004) reported that seepage increased by more than one order of magnitude for several seconds when an observer walked within 1 meter of a seepage chamber installed in a sandy lakebed, but the seepage rate changed only slightly when averaged over a minute-long period. This source of error is most substantial for small rates of seepage and can be avoided if the observer floats in the water while attaching and detaching the bag. Servicing the meter from a small boat or raft (Harvey and others, 2000) works well if the meter can be reached from the water surface. Using a short piece of hose or tubing to locate the bag 1 to 2 meters away from the seepage chamber also minimizes this source of error.

Figure 16.   Plastic bag attached to a garden-hose shut-off valve. Bag is filled with a known volume of water and then purged of air. Valve is closed. Bag is threaded onto male threads on seepage meter, and valve then is opened to begin seepage measurement. (Photograph by Donald Rosenberry, U.S. Geological Survey.)

flow is from surface water to ground water, the velocity-head effect will reduce the loss of water from the seepage bag. Libelo and MacIntyre (1994) indicated that water flowing at a velocity of 0.2 meter per second or faster past an uncovered bag resulted in larger rates of seepage than for a bag placed inside a protective cover, and that the velocity-head effect could more than double the measured seepage flux. They indicated this type of error also could result from near-shore waves or currents. Murdoch and Kelly (2003) quantified the velocity-head effect and indicated it becomes substantial when the velocity of the moving water is 0.1 meter per second or greater. Both studies indicated that the velocity-head effect is proportional to the square of the surface-water velocity, consistent with equation 5. Landon and others (2001) used bag shelters for their seepage measurements in the Platte River in Nebraska. Sebestyen and Schnieder (2001) used a plastic shield to protect their seepage-meter bags in a lake in New York. Asbury (1990) noted that bags exposed to small currents could be pulled to the side by the current, folding the bag over the opening and closing off the tubing to which the bag was attached. This was an especially important problem for flow out of the bag and when the bag was nearly empty of water. Conversely, Cable and others (1997a) indicated that currents were not a problem for exposed seepage bags attached to meters installed in near-shore regions of the Florida Gulf coast, as long as windspeed was less than 15 knots.

A seepage chamber installed in moving water also may affect actual seepage rates in the vicinity of the meter. Shinn and others (2002) reported that measured seepage was always from ground water to surface water at their study sites near the Florida Keys, even during intervals when piezometer nests indicated reversals in the hydraulic-head gradient in response to tidal influences. They attributed this phenomenon to the effect of the seepage chamber extending into the flow field where ocean currents were relatively strong, which would cause water to be advected through the sediment beneath the seepage chamber and into the seepage bag. Huettel and others (1996) indicated that this process also occurs naturally where an uneven sediment bed (ripples, dunes) projects into a moving-water flow field. On the upstream side of the obstruction to flow (a dune or, in this instance, a seepage chamber), water velocity and, therefore, velocity head decreases, pressure head increases, and the pressure gradient drives flow into the sediments, beneath the rim of the chamber, and into the chamber. The same process occurs at the downstream side of the obstruction where an eddy forms to decrease velocity head and increase pressure head. Others who have made seepage measurements in marine settings indicated that the seepage cylinder is little affected by waves and currents (Corbett and Cable, 2003). Regardless of the net effect, the local flow field

If currents are present, seepage-meter measurements can be erroneously large or small, depending on the seepage direction. This is the result of velocity head associated with moving water,

$$h_v = \frac{v^2}{2g},$$ (5)

where

v is velocity of water flowing past the seepage bag ($L\ T^{-1}$),

and

g is acceleration due to gravity ($L\ T^{-2}$).

Velocity head is one component of the total hydraulic head in a stream, which is the sum of velocity head, pressure head, and elevation head. Velocity head inside a seepage bag is zero because water is not moving appreciably inside the bag. Because the flexible plastic bag can easily respond to any pressure gradients across the bag surface, the pressure inside the bag is the same as outside of the bag. Therefore, the total hydraulic head inside the bag is equal to the hydraulic head outside of the bag minus the velocity-head component. If flow is from ground water to surface water, the velocity-head effect will induce additional water to flow into the seepage bag. If

is undoubtedly altered by the presence of a seepage chamber positioned in a stream or river. Landon and others (2001) and Zamora (2006) reported scouring of the bed at several seepage-meter installations in a sand-bed river.

Leaks

A hole in the seepage meter bag is one of the most common types of leaks. This can be prevented by careful handling and frequent testing of the bag, and by "double bagging" the bag where fish or crustaceans may make holes in the bag. As mentioned previously, if bags are "double bagged," small holes should be placed in the corners of the outer bag to allow the evacuation of air trapped between the bags. Bag shelters also can minimize the potential for bag damage. The attachment between the bag and the device that connects the bag to the chamber is another potential location for leaks. The attachment method may involve electrical tape, rubber bands, or plastic cable ties, and care should be used to ensure a good connection to the tubing or other mechanical connector. One solution to this potential problem is to use a bag manufactured with an integral plastic tube or sleeve. As previously mentioned, however, many of these bags typically are much thicker than food-storage bags and likely resist movement in response to changing fluid volume inside the bag. One bag that shows promise is designed for use as shipping-protection material. The bag is designed to be inflated through a plastic neck that is manufactured as part of the bag. A tube is inserted through the neck and sealed with adhesive or electrical tape. The bag material is quite thin (25 micrometers), and tests conducted by Murdoch and Kelly (2003) and Rosenberry and Menheer (2006) indicate that it presents little resistance to filling.

Leaks also can occur in the seepage chamber because of rust, improper welds, or improper sealing where the tubing passes through the rubber stopper (if that is the mechanism used) or between the rubber stopper and the chamber. If the seepage chamber contains a bung, then a loose bung or a weathered or cracked bung gasket also can lead to leaks.

Measurement Error

The change in the volume of water in the seepage-meter bag commonly is measured by use of a graduated cylinder. Sources for error include misreading the meniscus on the graduated cylinder, not holding the graduated cylinder level when making a reading, not removing all of the water from inside the bag, spilling water during filling or emptying of the bag, and misrecording time of attachment and time of removal. A funnel is useful for eliminating spills during filling and emptying the bag. Another method of measuring volume change involves weighing the bag with an accurate, portable electronic scale before bag attachment and again following bag removal. Additionally, to reduce the uncertainty of volume-measurement error, many investigators commonly make three or more measurements at each site and average the values.

Flexible Seepage-Meter Chamber

Occasionally the flat, circular end of a half-barrel seepage meter can flex downward or upward (sometimes with a sudden, audible pop) in response to temperature changes or to pressure applied to the metal surface. Standing on the center of the meter during emplacement, for example, can cause such flexing. If the metal later returns to a more relaxed position while the bag is attached, an erroneous measurement will result. Allowing time for equilibration between installation and first measurement minimizes the likelihood of this occurring during subsequent measurements. Other types of chambers also may have insufficient rigidity. Plastic trash cans can flex if the walls of the plastic are too thin. Shinn and others (2002) constructed a seepage meter with a flexible top with the intent that the meter would flex with the passage of waves; associated pressure perturbations exerted on the ocean bed also would be exerted on the part of the bed covered by the meter. This was done to reduce water artificially advected into the meter; the experiment met with little success.

Insufficient or Excessive Bag-Attachment Time

Bags need to be attached to a seepage meter long enough for a measurable change in volume in the bag to occur, but not so long that the bag is either full or empty. Bag-attachment times can range from seconds to weeks, depending on the size of the bag, the diameter of the seepage chamber, and the rate of seepage. Problems related to insufficient or excessive attachment time are obvious when the bag is full or empty upon removal of the bag. A bag that is nearly full or nearly empty when being removed also may indicate an erroneous flux rate. As mentioned earlier, Murdoch and Kelly (2003) determined that the head required to move water into a bag increases markedly when the bag is within a few hundred milliliters of being full. Such a condition also is likely when the bag is losing water and approaches being empty. Conversely, the bag may contain nearly the same volume of water following removal as it contained during attachment, indicating that the bag-attachment time was too short. The solution to both problems is an iterative one. Subsequent measurement periods can be adjusted based on previous incorrect attachment times.

Accumulation of Trapped Gas

Release of gas from sediments is common where organic decomposition produces methane, carbon dioxide, hydrogen sulfide, or other gases. If gas accumulates within the seepage chamber, it can displace water from inside the chamber that then is forced into the bag. There are at least two solutions to this problem. A vent tube may be installed at the highest point of the meter that extends above the water surface to the atmosphere (fig. 13.4, lower panel). This allows gas released from the sediments covered by the chamber to be released to the atmosphere instead of accumulating within the chamber. Alternatively, gas may be allowed to escape to the bag. Boyle (1994) designed a meter that automatically allowed gas to escape from the meter without being transmitted to the bag. Hedblom and others (2003) described a system designed to collect both gas and water released from contaminated sediments; they analyzed both gas and water to determine the rates of release of various chemicals.

Use of Improper Correction Coefficient

Numerous tests have been conducted to compare flow through seepage meters with the rate of seepage in a controlled-flow test tank (Lee, 1977; Erickson, 1981; Asbury, 1990; Cherkauer and McBride, 1988; Dorrance, 1989; Belanger and Montgomery, 1992; Murdoch and Kelly, 2003; Rosenberry, 2005; Rosenberry and Menheer, 2006). Results of these tests indicate that seepage meters undermeasure the flux of water across the sediment-water interface because of frictional flow loss within the meter, restrictions to flow through the connector between the bag and the chamber, and any resistance to movement of the bag. Coefficients typically are applied to the indicated flux to correct for this problem. Erickson (1981) determined that the coefficient was different depending on the direction of flow. His studies indicated a multiplier of 1.43 was required for flow from ground water to surface water and a multiplier of 1.74 for flow from surface water to ground water. Belanger and Montgomery (1992) indicated a multiplier of 1.30 was required to correct for measurements of flow from ground water to surface water. Cherkauer and McBride (1988) used a correction factor of 1.6 for flow from ground water to surface water, and Dorrance (1989) indicated that a multiplier of 1.61 was required for his seepage meter designed for quantifying loss of water from a reservoir. Asbury (1990) used a multiplier for flow either into or out of a surface-water body of 1.11; he attributed his lower multiplier to his using a larger diameter connector (19 millimeters) than other investigators. Murdoch and Kelly (2003) determined measurement inefficiency by using highly accurate manometers to measure head loss, and reported correction factors of 1.25 to 1.82, depending on the type of bag used. Rosenberry (2005) used large diameter (9.5-millimeter minimum inside diameter) connection materials and a thin-walled 4-liter bag with a Lee-type seepage chamber to obtain a correction factor of 1.05.

Fellows and Brezonik (1980) related seepage-meter efficiency to the diameter of the connector between the meter and the bag and to seepage velocity. Their results indicated that head loss increased with decreasing tubing diameter and with increasing seepage velocity (fig. 17); they suggested that a tubing diameter larger than 5 millimeters would not cause loss of efficiency for most fluxes commonly measured with seepage meters. On the basis of their experiments, however, they altered their seepage-meter design to use a 9-millimeter opening instead of a 5-millimeter opening between the bag and the meter. Rosenberry and Morin (2004) found a similar response by positioning a pressure transducer inside a seepage meter and recording pressure changes in response to routing seepage through a range of tubing diameters. Pressure changed by 21 millimeters of water head when seepage was forced to flow

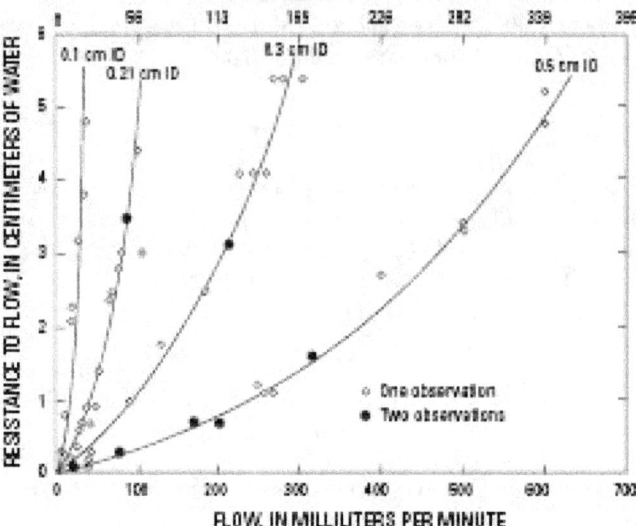

Figure 17. Resistance to flow related to tubing diameter and rate of seepage. Seepage flux assumes a 0.25-square-meter-area seepage meter. (Modified from Fellows and Brezonik, 1980; copyright 1980 by the American Water Resources Association, used with permission.)

through 4-millimeter-diameter tubing, but the pressure change was only 4 millimeters of water head when flow was routed through 7.9-millimeter-diameter tubing. Harvey and others (2000) used a large-diameter (19-millimeter) connection system to eliminate any concern regarding tubing resistance in a study of seepage from wetlands in the Florida Everglades. Rosenberry and Menheer (2006) describe a seepage-meter calibration tank for determining the efficiency of various seepage-meter designs.

Insufficient Characterization of Spatial Heterogeneity in Seepage Through Sediments

Successful extrapolation of point measurements of seepage to whole-lake systems requires that the seepage measurements adequately characterize the larger scale integrated exchange between ground water and surface water. This extrapolation can be difficult because small-scale spatial variability in flux across the sediment-water interface is common. Measurements at several locations may be required to adequately characterize seepage on a meaningful spatial scale. Shaw and Prepas (1990a) determined that seepage rates could vary by more than a factor of 2 when meters were installed only 1 meter apart (fig. 18). They found that seepage flux in a 2-square-meter area was lognormally distributed, and the variance in seepage increased with seepage velocity. They attributed seepage variability to variability in hydraulic conductivity of the lakebed. Shaw and Prepas (1990b) recommended making seepage measurements at additional transects in a lake rather than making replicate measurements at a single transect to best characterize spatial variability in lakebed seepage.

Asbury (1990) addressed the question of seepage-meter precision related to lakebed heterogeneity by installing 25 seepage meters on an 8-meter by 8-meter grid. His results (table 3) showed a large decrease in seepage with distance from shore and then a reversal in seepage direction farther from shore as was expected based on previous results. The five measurements made at each distance from shore showed remarkable consistency near the shoreline where seepage rates were largest, but seepage variability increased with distance from shore out to 6 meters from shore. Beyond that distance, seepage direction reversed and the variance decreased slightly.

Belanger and Walker (1990) tested small-scale spatial variability in seepage by placing two to three seepage meters 5 meters apart at seven different sites. They found very good reproducibility at five of the sites where seepage rates were relatively small. At the other two sites, where seepage rates were much larger, they attributed the greater spatial variability in seepage to the presence of springs in the area.

Michael and others (2003) used 40 seepage meters to measure seepage variability in four transects perpendicular from shore on a 50-meter spacing in a saltwater bay near Cape Cod, Massachusetts. They detected bands of seepage with distance from shore that were parallel to the shoreline and determined that as long as meters are arranged in transects, errors associated with reducing the number of transects are not unacceptably large. Departures from flux estimated with all four transects were 9, 4, and 3 percent when data from one, two, or three transects were used. They also placed seepage meters in clusters with 1-meter spacing and found spatial variability in seepage of the same magnitude as with the 50-meter spacing.

Approaches to characterizing seepage variability include either making numerous measurements in each area of interest, in a manner similar to the approach of Asbury (1990) or Michael and others (2003), or using larger seepage chambers that cover larger areas of the sediment-water interface and better integrate the heterogeneity in seepage. Typically, the scale of interest is a characterization of seepage for an entire surface-water body or shoreline reach. In this instance, resources may be better spent characterizing seepage along a number of transects positioned throughout the area of interest, which characterizes spatial variability on a scale appropriate for the interests of the study (for example, Michael and others, 2003). Rosenberry (2005) addressed the heterogeneity issue by routing flow from several seepage chambers to one collection bag. With such a system, spatial variability in seepage is averaged in one measurement, which also reduces bag-collection time and labor costs. Head loss did not substantially reduce the efficiency of the ganged seepage measurement when 3-meter lengths of garden hose (14-millimeter diameter) were used to connect the seepage chambers.

Best-Measurement Practices for Manual Seepage Meters

The following recommendations are presented for minimizing errors associated with making seepage-meter measurements:

1. Use a rigid seepage chamber. A diameter of approximately 0.5 meter seems to be a useful compromise between maximizing areal coverage and maximizing convenience of use. Make certain that the entire rim of the seepage chamber is seated at least a few centimeters into the sediment-water interface. For sandy sediments, 1 hour is probably a sufficient time to wait between installation and first bag measurements. For softer sediments, it may be prudent to wait 1 day to begin measurements.

2. Use several meters to characterize spatial heterogeneity at a scale that is appropriate for the interests of the study. Seepage chambers can be ganged to integrate seepage heterogeneity over a larger area and also to minimize the number of required bag measurements.

3. Use a shelter to protect the bag from waves and currents and to ensure that the bag orientation is maintained in a position that will not close or restrict the opening between the bag and the bag-connection system.

4. Use a large-diameter bag-connection system, especially when fast seepage rates are expected. A diameter 9 millimeters or larger is suggested.

5. Use thin-walled bags to minimize bag resistance. A bag size of 4 liters is convenient for most seepage rates.

6. Prefill the bag with 500 to 1,000 milliliters of water prior to bag attachment. If seepage from surface water to ground water is expected, a larger initial volume of water may be warranted. Do not fill the bag to more than about 75 percent of its capacity.

7. Seepage-meter correction coefficients have been decreasing over time as seepage-meter designs become more efficient. If the suggestions listed above are followed, a coefficient from 1 to 1.1 will provide a good estimate of true seepage rates for most meter designs.

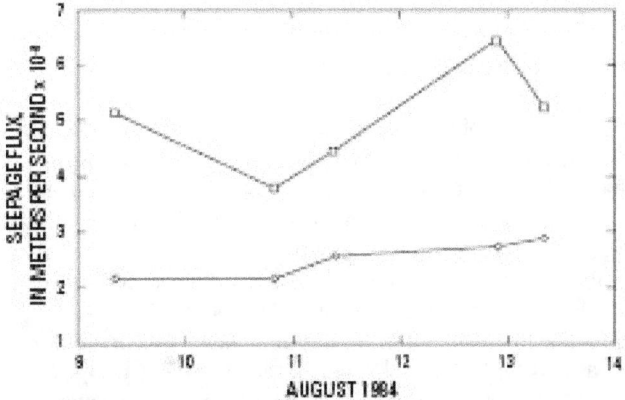

Figure 18. Seepage flux measured at two seepage meters located 1 meter apart. Flux values are in meters per second. (Modified from Shaw and Prepas, 1990a; copyright 1990, reprinted from Journal of Hydrology, used with permission from Elsevier.)

Table 3.   Seepage flux with distance from shore and distance along shore on an 8-meter by 8-meter grid (2-meter seepage-meter spacing) (from Asbury, 1990).

Distance from shore (meters)	Seepage flux (centimeters per day)						
	Distance along shore (meters)					Average	Variance
	0	2	4	6	8		
2	−11.3	−11.2	−11.4	−12.1	−11.8	−11.56	0.143
4	−8.4	−9.3	−10.4	−10.5	−6.5	−9.02	2.727
6	−3.5	−6.7	−6.4	−0.8	−4.5	−4.38	5.767
8	0.9	2.4	3.8	3.1	4.9	3.02	2.257
10	3.7	0.3	2.3	0.2	1.2	1.54	2.173

Automated Seepage Devices

Temporal variability in flux across the interface between ground water and surface water has been investigated on a seasonal scale (for example, Schneider and others, 2005; Michael and others, 2005), but temporal variability on a weekly or shorter time scale has not been extensively investigated. Several investigators have made numerous measurements over time to measure the temporal variability (Lee, 1977; Cable and others, 1997b; Sebestyen and Schneider, 2001), but this is a labor-intensive endeavor. Recently developed automated devices allow measurement of seepage responses to temporal events such as seiches (Taniguchi and Fukuo, 1996), tides (Paulsen and others, 2001, 2004; Taniguchi, 2002), and recharge events (Rosenberry and Morin, 2004).

Several of these automated devices use heat-pulse technology to measure flow. One such meter uses sensors originally developed for measuring sap flow in plants (Taniguchi and Fukuo, 1993, 1996) and records the data with a digital datalogger enclosed in the submerged seepage meter. Another design uses the same heat-pulse technology but also includes sensors for collection of water-quality data (Krupa and others, 1998). This device is tethered to a raft that is anchored above the submerged seepage meter. Taniguchi has recently improved the heat-pulse method with a continuous heat-source seepage meter (Taniguchi and others, 2003).

Paulsen and others (2001) developed an automated seepage meter that makes use of acoustic-velocity technology more commonly used to measure surface-water flow. Their sensor can measure flux velocity values ranging from about 1 to at least 275 centimeters per day over an exchange area of 0.21 square meter. Menheer (2004) also used an acoustic-velocity sensor to measure seepage in a benthic-flux chamber that was designed to quantify flow of mercury from ground water to surface water. Another automated seepage meter replaces the plastic bag with an electromagnetic flow meter typically used to measure flow velocity in boreholes (Rosenberry and Morin, 2004). With the flow meter attached to a 1.1-meter-diameter chamber, a modified version of their sensor can measure flux velocities ranging from 4 to 4,000 centimeters per day.

A device developed for use in deep-ocean environments uses a chemical tracer that is injected into an outlet tube of the seepage meter (Tyron and others 2001). A pair of sample-collection coils on either side of the injection point provides a record of the tracer based on dilution of the injectate relative to the seepage rate. Water in the coils is sampled upon retrieval of the meter and analyzed to provide time-series data of the seepage rate. This device can measure seepage rates ranging from 3×10^{-5} to 4 centimeters per day.

Dye-dilution seepage meters make use of dye-dilution chambers, the size of which can be adjusted to accommodate a wide range of seepage rates (Sholkovitz and others, 2003). The combination of chambers used with the meter developed by Sholkovitz and others can measure seepage rates ranging from less than 0.1 to more than 300 centimeters per day. The authors point out that smaller chambers could be used to measure smaller seepage rates at deep-ocean installations.

The Taniguchi and Krupa automated seepage meters have been in use for 10 to 15 years, but as of 2007, the other meters are in the early stages of use. These automated devices are not subject to the previously mentioned problems associated with the use of seepage bags. Although all of the automated devices are fitted to a seepage chamber and are subject to the chamber- and connection-related errors discussed above, those errors should be relatively small compared to bag-related errors.

Methods Selection

Selection of the appropriate methods of calculation and (or) measurement is one of the most important decisions to be made when quantifying exchange between ground water and surface water. Of the three methods presented in this chapter, each has advantages and disadvantages that may or may not be relevant to the study area of interest. Although it is not possible to anticipate all situations, table 4 provides a general guideline to conditions or situations in which each method is particularly well- or ill-suited. As indicated in Chapter 1, the use of more than one method to quantify the exchange between ground water and surface water can be informative and valuable to increasing the confidence in the flux values estimated or calculated.

Table 4.   Conditions for which methods for quantifying flow between ground water and surface water are well- or ill-suited.

Method	Well-suited for:	Ill-suited for:
Calculations from water levels in network of wells and surface-water stage	• Basin-scale quantification • Distinguishing areas of inflow from areas of outflow • Determining large-scale aquifer characteristics • Relatively homogeneous aquifers	• Determining flux of some chemicals that enter or leave a surface-water body • Steep and (or) rocky shorelines where installation of wells is difficult or impossible • Low-lying terrain where shoreline migration is large and evapotranspiration is a significant factor • Areas with complex geology or vertical flow regimes where effective depth of aquifer is nearly impossible to determine
Hydraulic potentiomanometer and well-probe measurements	• Fine sand to medium gravel sediments • Quick reconnaissance for qualitative determination of direction of flow • Determining variability of vertical hydraulic gradient with depth • Collection of water-quality samples	• Fine-grained sediments • Rocky shorelines or bedrock • Surface-water body with any appreciable wave action • Fast-flowing water • Organic, gas-rich sediments
Seepage-meter measurements	• Direct measurement of seepage flux • Areal distribution of seepage flux • Sediments ranging from clayey-silt to fine-medium gravel • Calm-water settings • Shallow-water settings	• Surface-water body with any appreciable wave action • Areas with strong currents or fast-flowing water • Very soft, low-density sediments • Rocky sediment beds • Bed areas with dense vegetation

References

Asbury, C.E., 1990, The role of groundwater seepage in sediment chemistry and nutrient budgets in Mirror Lake, New Hampshire: Cornell University, Ph.D., 275 p.

Belanger, T.V., and Kirkner, R.A., 1994, Ground-water/surface-water interaction in a Florida augmentation lake: Lake and Reservoir Management, v. 8, no. 2, p. 165–174.

Belanger, T.V., and Mikutel, D.F., 1985, On the use of seepage meters to estimate groundwater nutrient loading to lakes: Water Resources Bulletin, v. 21, no. 2, p. 265–272.

Belanger, T.V., and Montgomery, M.T., 1992, Seepage meter errors: Limnology and Oceanography, v. 37, no. 8, p. 1787–1795.

Belanger, T.V., and Walker, R.B., 1990, Ground water seepage in the Indian River Lagoon, Florida: American Water Resources Association Technical Publication Series TPS, v. 90–2, p. 367–375.

Blanchfield, P.J., and Ridgway, M.S., 1996, Use of seepage meters to measure groundwater flow at brook trout redds: Transactions of the American Fisheries Society, v. 125, p. 813–818.

Boyle, D.R., 1994, Design and seepage meter for measuring groundwater fluxes in the nonlittoral zones of lakes—Evaluation in a boreal forest lake: Limnology and Oceanography, v. 39, no. 3, p. 670–681.

Brock, T.D., Lee, D.R., Janes, D., and Winek, D., 1982, Groundwater seepage as a nutrient source to a drainage lake: Lake Mendota, Wisconsin: Water Resources, v. 16, p. 1255–1263.

Brunke, M., 1999, Colmation and depth filtration within streambeds—Retention of particles in hyporheic interstices: International Review of Hydrobiology, v. 84, no. 2, p. 99–117.

Buchanan, T.J., and Somers, W.P., 1982, Stage measurement at gaging stations: U.S. Geological Survey Techniques of Water-Resources Investigations, book 3, chap. A7, 28 p.

Cable, J.E., Burnett, W.C., Chanton, J.P., Corbett, D.R., and Cable, P.H., 1997a, Field evaluation of seepage meters in the coastal marine environment: Estuarine, Coastal and Shelf Science, v. 45, p. 367–375.

Cable, J.E., Burnett, W.C., and Chanton, J.P., 1997b, Magnitude and variations of groundwater seepage along a Florida marine shoreline: Biogeochemistry, v. 38, p. 189–205.

Carr, M.R., and Winter, T.C., 1980, An annotated bibliography of devices developed for direct measurement of seepage: U.S. Geological Survey Open-File Report 80–344, 38 p.

Cedergren, H.R., 1997, Seepage, drainage, and flow nets (3d ed.): New York, John Wiley and Sons, 465 p.

Cherkauer, D.A., and McBride, J.M., 1988, A remotely operated seepage meter for use in large lakes and rivers: Ground Water, v. 26, no. 2, p. 165–171.

Corbett, D.R., and Cable, J.E., 2003, Seepage meters and advective transport in coastal environments—Comments on "Seepage Meters and Bernoulli's Revenge" by E.A. Shinn, C.D. Reich, and T.D. Hickey, 2002, Estuaries 25:126–132: Estuaries, v. 26, no. 5, p. 1383–1389.

Davis, S.N., and DeWiest, R.J.M., 1991, Hydrogeology. Malabar, Florida: Krieger Publishing, 463 p.

Dorrance, D.W., 1989, Streaming potential and seepage meter studies at Upper Lake Mary near Flagstaff, Arizona: Tucson, University of Arizona, Masters thesis, 182 p.

Duff, J.H., Murphy, F., Fuller, C.C., Triska, F.J., Harvey, J.W., and Jackman, A.P., 1998, A mini drivepoint sampler for measuring pore water solute concentrations in the hyporheic zone of sand-bottom streams: Limnology and Oceanography, v. 43, no. 6, p. 1378–1383.

Duff, J.H., Toner, B., Jackman, A.P., Avanzino, R.J., and Triska, F.J., 1999, Determination of groundwater discharge into a sand and gravel bottom river—A comparison of chloride dilution and seepage meter techniques: Verh. Internat. Verein. Limnol., v. 27, p. 406–411.

Erickson, D.R., 1981, A study of littoral groundwater seepage at Williams Lake, Minnesota, using seepage meters and wells: Minneapolis, University of Minnesota, Master's thesis, 135 p.

Fellows, C.R., and Brezonik, P.L., 1980, Seepage flow into Florida lakes: Water Resources Bulletin, v. 16, no. 4, p. 635–641.

Fetter, C.W., 2000, Applied hydrogeology (4th ed.): Englewood Cliffs, N.J., Prentice Hall, 691 p.

Fleckenstein, J.H., Niswonger, R.G., and Fogg, G.E., 2006, River-aquifer interactions, geologic heterogeneity, and low-flow management: Ground Water, v. 44, no. 6, p. 837–852.

Harbaugh, A.W., 2005, MODFLOW-2005, the U.S. Geological Survey modular ground-water model—The ground-water flow process: U.S. Geological Survey Techniques and Methods 6–A16, 237 p.

Harbaugh, A.W., Banta, E.R., Hill, M.C., and McDonald, M.G., 2000, MODFLOW-2000, the U.S. Geological Survey modular ground-water model—User guide to modularization concepts and the ground-water flow process: U.S. Geological Survey Open-File Report 2000–92, 121 p.

Harvey, J.W., Krupa, S.L., Gefvert, C.J., Choi, J., Mooney, R.H., and Giddings, J.B., 2000, Interaction between ground water and surface water in the northern Everglades and relation to water budgets and mercury cycling—Study methods and appendixes: U.S. Geological Survey Open-File Report 2000–168, 395 p.

Harvey, J.W., Krupa, S.L., and Krest, J.M., 2004, Ground water recharge and discharge in the Central Everglades: Ground Water, v. 42, no. 7, p. 1090–1102.

Hedblom, E., Costello, M., and Huls, H., 2003, Integrated field sampling for design of a remedial cap: Cincinnati, Ohio, Proceedings of the In-Situ Contaminated Sediment Capping Workshop, May 12–14, 2003, p. 19.

Henry, M.A., 2000, Appendix D: MHE push-point sampling tools, in Proceedings of the Ground-Water/Surface-Water Interactions Workshop, July 2000: U.S. Environmental Protection Agency, EPA/542/R–00/007, p. 199–200.

Hiscock, K.M., and Grischek, T., 2002, Attenuation of ground-water pollution by bank filtration: Journal of Hydrology, v. 266, p. 139–144.

Huettel, M., Ziebis, W., and Forster, S., 1996, Flow-induced uptake of particulate matter in permeable sediments: Limnology and Oceanography, v. 41, no. 2, p. 309–322.

Hunt, R.J., Haitjema, H.M., Krohelski, J.T., and Feinstein, D.T., 2003, Simulating ground water—Lake interactions; approaches and insights: Ground Water, v. 41, no. 2, p. 227–237.

Hvorslev, M.J., 1951, Time lag and soil permeability in ground water observations: U.S. Army Corps of Engineers Waterways Experimental Station Bulletin no. 36, 50 p.

Isiorho, S.A., and Meyer, J.H., 1999, The effects of bag type and meter size on seepage meter measurements: Ground Water, v. 37, no. 3, p. 411–413.

Israelson, O.W., and Reeve, R.C., 1944, Canal lining experiments in the Delta Area, Utah: Utah Agricultural Experimental Station, Bulletin 313, p. 15–35.

Kelly, S.E., and Murdoch, L.C., 2003, Measuring the hydraulic conductivity of shallow submerged sediments: Ground Water, v. 41, no. 4, p. 431–439.

Kim, K., Anderson, M.P., and Bowser, C.J., 1999, Model calibration with multiple targets—A case study: Ground Water, v. 37, no. 3, p. 345–351.

Krabbenhoft, D.P., and Anderson, M.P., 1986, Use of a numerical ground-water flow model for hypothesis testing: Ground Water, v. 24, p. 49–55.

Krupa, S.L., Belanger, T.V., Heck, H.H., Brock, J.T., and Jones, B.J., 1998, Krupaseep—The next generation seepage meter, in International Coastal Symposium (ICS 98), Journal of Coastal Research Special Issue no. 26: Fort Lauderdale, Fla., Coastal Education Research Foundation, p. 210–213.

Landon, M.K., Rus, D.L., and Harvey, F.E., 2001, Comparison of instream methods for measuring hydraulic conductivity in sandy streambeds: Ground Water, v. 39, no. 6, p. 870–885.

Leake, Stanley A., 1997, Modeling ground-water flow with MODFLOW and related programs: U.S. Geological Survey Fact Sheet FS–121–97, 4 p.

Lee, D.R., 1977, A device for measuring seepage flux in lakes and estuaries: Limnology and Oceanography, v. 22, no. 1, p. 140–147.

Lee, D.R., and Cherry, J.A., 1978, A field exercise on ground-water flow using seepage meters and mini-piezometers: Journal of Geological Education, v. 27, p. 6–20.

Lee, T.M., and Swancar, A., 1997, Influence of evaporation, ground water, and uncertainty in the hydrologic budget of Lake Lucerne, a seepage lake in Polk County, Florida: U.S. Geological Survey Water Supply Paper 2439, 61 p.

Lewis, J.B., 1987, Measurements of groundwater seepage flux onto a coral reef—Spatial and temporal variations: Limnology and Oceanography, v. 32, no. 5, p. 1165–1169.

Libelo, E.L., and MacIntyre, W.G., 1994, Effects of surface-water movement on seepage-meter measurements of flow through the sediment-water interface: Applied Hydrogeology, v. 2, no. 4, p. 49–54.

Lock, M.A., and John, P.J., 1978, The measurement of groundwater discharge into a lake by direct method: Internationale Revue der gesamten Hydrobiologie, v. 63, no. 2, p. 271–275.

Lundy, J.R., and Ferrey, Mark, 2004, Direct measurement of ground water contaminant discharge to surface water, in Proceedings of National Water Quality Monitoring Council 2004 National Monitoring Conference, May 17–20, 2004: Chattanooga, Tenn., 10 p.

McCobb, T.D., LeBlanc, D.R., and Socolow, R.S., 1999, A siphon gage for monitoring surface-water levels: Journal of the American Water Resources Association, v. 35, no. 5, p. 1141–1146.

Menheer, M.A., 2004, Development of a benthic-flux chamber for measurement of ground-water seepage and water sampling for mercury analysis at the sediment-water interface: U.S. Geological Survey Scientific Investigations Report 2004–5298, 14 p.

Michael, H.A., Lubetsky, J.S., and Harvey, C.F., 2003, Characterizing submarine groundwater discharge—A seepage meter study in Waquoit Bay, Massachusetts: Geophysical Research Letters, v. 30, no. 6, p. 1297. [doi:10.1029/2002GL016000]

Michael, H.A., Mulligan, A.E., and Harvey, C.F., 2005, Seasonal oscillations in water exchange between aquifers and the coastal ocean: Nature, v. 436, p. 1145–1148.

Mitchell, D.F., Wagner, K.J., and Asbury, C., 1988, Direct measurement of groundwater flow and quality as a lake management tool: Lake and Reservoir Management, v. 4, no. 1, p. 169–178.

Murdoch, L.C., and Kelly, S.E., 2003, Factors affecting the performance of conventional seepage meters: Water Resources Research, v. 39, no. 6, p. SWC 2–1. [doi:10.1029/2002WR001347]

Nield, S.P., Townley, L.R., and Barr, A.D., 1994, A framework for quantitative analysis of surface water-groundwater interaction—Flow geometry in a vertical section: Water Resources Research, v. 30, no. 8, p. 2461–2475.

Paulsen, R.J., Smith, C.F., O'Rourke, D., and Wong, T., 2001, Development and evaluation of an ultrasonic ground water seepage meter: Ground Water, v. 39, no. 6, p. 904–911.

Paulsen, R.J., O'Rourke, D., Smith, C.F., and Wong, T.-F., 2004, Tidal load and salt water influences on submarine ground water discharge: Ground Water, v. 42, no. 7, p. 990–999.

Rasmussen, W., and Lauritzen, C.W., 1953, Measuring seepage from irrigation canals: Agricultural Engineering, v. 34, p. 326–330.

Robinson, A.R., and Rohwer, C., 1952, Study of seepage losses from irrigation channels: U.S. Department of Agriculture, Soil Conservation Service, Progress Report, 42 p.

Rosenberry, D.O., 2000, Unsaturated-zone wedge beneath a large, natural lake: Water Resources Research, v. 36, no. 12, p. 3401–3409.

Rosenberry, D.O., 2005, Integrating seepage heterogeneity with the use of ganged seepage meters: Limnology and Oceanography, Methods, v. 3, p. 131–142.

Rosenberry, D.O., and Menheer, M.A., 2006, A system for calibrating seepage meters used to measure flow between ground water and surface water: U.S. Geological Survey Scientific Investigations Report 2006–5053, 21 p.

Rosenberry, D.O., and Morin, R.G., 2004, Use of an electromagnetic seepage meter to investigate temporal variability in lake seepage: Ground Water, v. 42, no. 1, p. 68–77.

Rovey, C.W., II, and Cherkauer, D.S., 1995, Scale dependency of hydraulic conductivity measurements: Ground Water, v. 33, no. 5, p. 769–780.

Schincariol, R.A., and McNeil, J.D., 2002, Errors with small volume elastic seepage meter bags: Ground Water, v. 40, no. 6, p. 649–651.

Schneider, R.L., Negley, T.L., and Wafer, C., 2005, Factors influencing groundwater seepage in a large, mesotrophic lake in New York: Journal of Hydrology, v. 310, p. 1–16.

Schubert, J., 2002, Hydraulic aspects of riverbank filtration—Field studies: Journal of Hydrology, v. 266, p. 145–161.

Schulze-Makuch, D., Carlson, D.A., Cherkauer, D.S., and Malik, P., 1999, Scale dependency of hydraulic conductivity in heterogeneous media: Ground Water, v. 37, no. 6, p. 904–919.

Sebestyen, S.D., and Schneider, R.L., 2001, Dynamic temporal patterns of nearshore seepage flux in a headwater Adirondack lake: Journal of Hydrology, v. 247, p. 137–150.

Shaw, R.D., and Prepas, E.E., 1989, Anomalous, short-term influx of water into seepage meters: Limnology and Oceanography, v. 34, no. 7, p. 1343–1351.

Shaw, R.D., and Prepas, E.E., 1990a, Groundwater-lake interactions, I—Accuracy of seepage meter estimates of lake seepage: Journal of Hydrology, v. 119, p. 105–120.

Shaw, R.D., and Prepas, E.E., 1990b, Groundwater-lake interactions, II—Nearshore seepage patterns and the contribution of groundwater to lakes in central Alberta: Journal of Hydrology, v. 119, p. 121–136.

Sheets, R.A., Darner, R.A., and Whitteberry, B.L., 2002, Lag times of bank filtration at a well field, Cincinnati, Ohio, USA: Journal of Hydrology, v. 266, p. 162–174.

Shinn, E.A., Reich, C.D., and Hickey, T.D., 2002, Seepage meters and Bernoulli's Revenge: Estuaries, v. 25, no. 1, p. 126–132.

Sholkovitz, E., Herbold, C., and Charette, M., 2003, An automated dye-dilution based seepage meter for the time-series measurement of submarine groundwater discharge: Limnology and Oceanography—Methods, v. 1, p. 16–28.

Siegel, D.I., and Winter, T.C., 1980, Hydrologic setting of Williams Lake, Hubbard County, Minnesota: U.S. Geological Survey Open-File Report 80-403, 56 p.

Squillace, P.J., Liszewski, M.J., and Thurman, E.M., 1993, Agricultural chemical interchange between ground water and surface water, Cedar River basin, Iowa and Minnesota—A study description: U.S. Geological Survey Open-File Report 92-0085, 26 p.

Taniguchi, M., 2001, Evaluation of the groundwater capture zone for modelling of nutrient discharge: Hydrological Processes, v. 15, p. 1939–1949.

Taniguchi, M., 2002, Tidal effects on submarine groundwater discharge into the ocean: Geophysical Research Letters, v. 29, no. 12, p. 2–1. [doi:10.1029/2002GL014987]

Taniguchi, M., Burnett, W.C., Smith, C.F., Paulsen, R.J., O'Rourke, D., Krupa, S.L., and Christoff, J.L., 2003, Spatial and temporal distributions of submarine groundwater discharge rates obtained from various types of seepage meters at a site in the Northeastern Gulf of Mexico: Biogeochemistry, v. 66, p. 35–53.

Taniguchi, M., and Fukuo, Y., 1993, Continuous measurements of ground-water seepage using an automatic seepage meter: Ground Water, v. 31, no. 4, p. 675–679.

Taniguchi, M., and Fukuo, Y., 1996, An effect of seiche on groundwater seepage rate into Lake Biwa, Japan: Water Resources Research, v. 32, no. 2, p. 333–338.

Townley, L.R., and Davidson, M.R., 1988, Definition of a capture zone for shallow water table lakes: Journal of Hydrology, v. 104, p. 53–76.

Townley, L.R., and Trefry, M.G., 2000, Surface water-groundwater interaction near shallow circular lakes—Flow geometry in three dimensions: Water Resources Research, v. 36, no. 4, p. 935–949.

Tyron, M., Brown, K., Dorman, L., and Sauter, A., 2001, A new benthic aqueous flux meter for very low to moderate discharge rates: Deep-Sea Research, v. 48, pt. I, p. 2121–2146.

Wanty, R.B., and Winter, T.C., 2000, A simple device for measuring differences in hydraulic head between surface water and shallow ground water: U.S. Geological Survey Fact Sheet FS–077–00, 2 p.

Warnick, C.C., 1951, Methods of measuring seepage loss in irrigation canals: University of Idaho, Engineering Experiment Station, Bulletin no. 8, 42 p.

Winter, T.C., LaBaugh, J.W., and Rosenberry, D.O., 1988, The design and use of a hydraulic potentiomanometer for direct measurement of differences in hydraulic head between groundwater and surface water: Limnology and Oceanography, v. 33, no. 5, p. 1209–1214.

Yelverton, G.F., and Hackney, C.T., 1986, Flux of dissolved organic carbon and pore water through the substrate of a Spartina alterniflora marsh in North Carolina: Estuarine, Coastal and Shelf Science, v. 22, p. 255–267.

Zamora, C., 2006, Estimates of vertical flux across the sediment—Water interface by direct measurement and using temperature as a tracer in the Merced River, California: Sacramento, California State University Sacramento, Master's thesis, 90 p.

Hydrogeologic Characterization and Methods Used in the Investigation of Karst Hydrology

By Charles J. Taylor and Earl A. Greene

Chapter 3 of
Field Techniques for Estimating Water Fluxes Between Surface Water and Ground Water
Edited by Donald O. Rosenberry and James W. LaBaugh

Techniques and Methods 4–D2

U.S. Department of the Interior
U.S. Geological Survey

Contents

Figures

Tables

Chapter 3
Hydrogeologic Characterization and Methods Used in the Investigation of Karst Hydrology

By Charles J. Taylor and Earl A. Greene

Introduction

Recharge to and discharge from ground water can be measured or estimated over a wide range of spatial and temporal scales in any hydrogeologic setting (National Academy of Sciences, 2004). Difficulties often arise in making these measurements or estimates because of insufficient knowledge of the processes involved in the transfer of water fluxes, inadequate characterization of the hydrogeologic framework in which they occur, and uncertainties in the measurements or estimates themselves. These difficulties may be magnified considerably in complex hydrogeological settings such as karst.

Karst is a unique hydrogeologic terrane in which the surface water and ground water regimes are highly interconnected and often constitute a single, dynamic flow system (White, 1993). The presence of karst usually is indicated by the occurrence of distinctive physiographic features that develop as a result of the dissolution of soluble bedrock such as limestone or dolostone (Field, 2002a). In well-developed karst, these physiographic features may include sinkholes, sinking (or disappearing) streams, caves, and karst springs. The hydrologic characteristics associated with the presence of karst also are distinctive and generally include: (1) internal drainage of surface runoff through sinkholes; (2) underground diversion or partial subsurface piracy of surface streams (that is sinking streams and losing streams); (3) temporary storage of ground water within a shallow, perched *epikarst* zone; (4) rapid, turbulent flow through subsurface pipelike or channellike solutional openings called *conduits*; and (5) discharge of subsurface water from conduits by way of one or more large perennial springs (fig. 1).

A karst aquifer can be conceptualized as an open hydrologic system having a variety of surface and subsurface input, throughput, and output flows, and boundaries defined by the catchment limits and geometry of conduits (Ford and Williams, 1989). The hydrogeologic characteristics of karst aquifers are largely controlled by the structure and organization of the conduits, the development of which generally acts to short-circuit surface drainage by providing alternative subsurface flow paths that have lower hydraulic gradients and resistance (White, 1999). Conduits are a third (tertiary) form of permeability that is distinctive from, yet interconnected with, the permeability provided by intergranular pores (bedrock matrix) and fractures. Because of the interconnection of matrix, fracture, and conduit permeability, karst aquifers are extremely heterogeneous compared to most granular and many fractured-rock aquifers and have hydraulic properties that are highly scale dependent and temporally variable (table 1).

Because of these unique hydrogeologic characteristics, data requirements for the hydrogeologic characterization of karst aquifers are somewhat more intensive and difficult to obtain than those for aquifers in most other types of hydrogeologic settings (Teutsch and Sauter, 1991). Wherever karst features are present, the water-resources investigator must anticipate the presence of a flow system that cannot be completely characterized by using conventional hydrogeologic methods such as potentiometric mapping or hydraulic tests of observation wells, by numerical modeling, or by using a study approach that treats ground water and surface water as separate hydrologic regimes (White, 1993). In karst terranes, a greater emphasis must generally be placed on the identification of hydrologic boundaries and subsurface flow paths, contributions of water from various recharge sources, and the structural and hydraulic properties of conduits. The acquisition of these data typically requires a multidisciplinary study approach that includes using more specialized investigation methods such as water-tracing tests and the analysis of variations in spring discharge and water chemistry (White, 1993; Ford and Williams, 1989).

This chapter presents an overview of methods that are commonly used in the hydrogeologic investigation and characterization of karst aquifers and in the study of water fluxes in karst terranes. Special emphasis is given to describing the techniques involved in conducting water-tracer tests using fluorescent dyes. Dye-tracer testing is a method successfully used in the study of karst aquifers in the United States and elsewhere for more than 30 years (Käss, 1998). However, dye-tracing techniques generally are not taught at the collegiate undergraduate or graduate level, lack a set of formalized peer-reviewed procedures, and sometimes are difficult to research because case studies often are reported in lesser-known publication venues outside the realm of mainstream professional journals (Beck, 2002). Dye-tracer test procedures described herein represent commonly accepted practices derived from a variety of published and previously unpublished sources. Methods that are commonly applied to the analysis of karst spring discharge (both flow and water chemistry) also are reviewed and summarized.

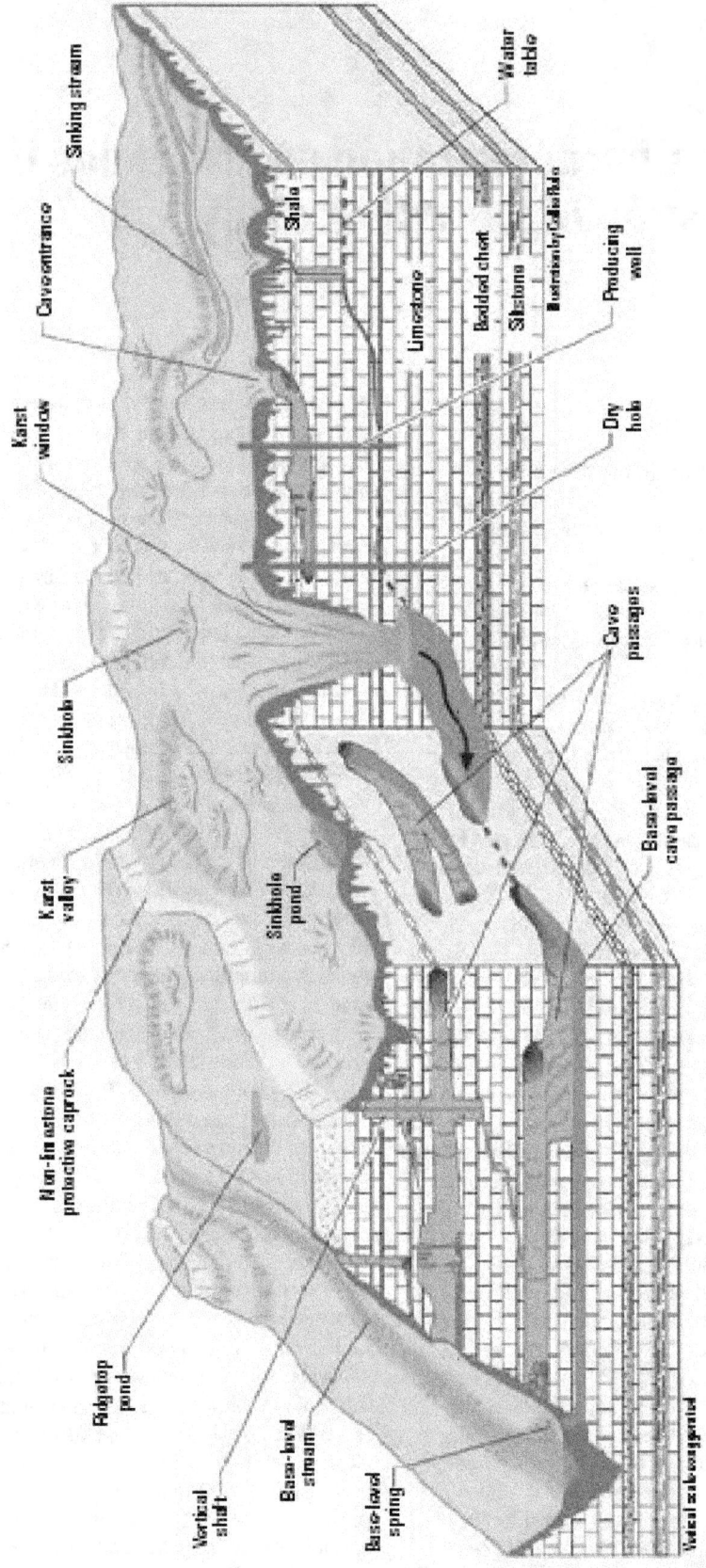

Figure 1.   Physiographic and hydrologic features typical of a well-developed karst terrane (modified from Currens, 2001, Kentucky Geological Survey, used with permission).

Table 1.   Comparison of various hydrogeologic properties for granular, fractured rock, and karst aquifers (ASTM, 2002).

Aquifer characteristics	Aquifer type		
	Granular	Fractured rock	Karst
Effective porosity	Mostly primary, through intergranular pores	Mostly secondary, through joints, fractures, and bedding plane partings	Mostly tertiary (secondary porosity modified by dissolution); through pores, bedding planes, fractures, conduits, and caves
Isotropy	More isotropic	Probably anisotropic	Highly anisotropic
Homogeneity	More homogeneous	Less homogeneous	Non-homogeneous
Flow	Slow, laminar	Possibly rapid and possibly turbulent	Likely rapid and likely turbulent
Flow predictions	Darcy's law usually applies	Darcy's law may not apply	Darcy's law rarely applies
Storage	Within saturated zone	Within saturated zone	Within both saturated zone and epikarst
Recharge	Dispersed	Primarily dispersed, with some point recharge	Ranges from almost completely dispersed- to almost completely point-recharge
Temporal head variation	Minimal variation	Moderate variation	Moderate to extreme variation
Temporal water chemistry variation	Minimal variation	Minimal to moderate variation	Moderate to extreme variation

Reprinted with permission from D 5717-95 Standard Guide for Design of Ground-Water Monitoring Systems in Karst and Fractured Rock Aquifers, copyright ASTM International, 100 Bar Harbor Drive, West Conshohocken, PA 19428.

Hydrogeologic Characteristics of Karst

A number of important characteristics of the physical hydrogeology of karst are summarized here for the benefit of readers less familiar with karst and with the differences between karst aquifers and aquifers in other hydrogeologic settings. The subject of karst hydrogeology involves a wide variety of geomorphologic, geologic, hydrologic, and geochemical topics that are beyond the scope of this report. White (1993, 1999) provides good overviews of karst hydrology and the methods typically used in its study. Other good sources of information about karst include textbooks written by Bogli (1980), White (1988), and Ford and Williams (1989); compendiums edited by Klimchouk and others (2000), and Culver and White (2004); and the proceedings of various karst conferences held in the United States from 1986 to 2005 (National Water Well Association, 1986, 1988; National Ground Water Association, 1991; Beck, 1995, 2003; Beck and Stephenson, 1997; Beck and others, 1999; and Kuniansky, 2001, 2002, 2005).

Many geological and hydrologic factors influence the development of karst, and not all karst features are present or developed to the same extent in every karst terrane. The information presented in this report best describes the hydrogeologic characteristics of fluviokarst and doline karst, which are common and widespread types of karst terranes in the United States (White, 1999). The term *fluviokarst* is used to describe a karst landscape in which the dominant physical landforms are valleys initially cut by surface streams that have been partly or completely diverted underground by subsurface conduit piracy (Field, 2002a). This type of karst is often typified by carbonate rocks that have low intrinsic permeability and is common of karst developed in Paleozoic limestones in the Interior Low Plateaus and Appalachian regions of the Eastern United States. The term *doline* karst describes karst landscape in which surface streams are almost entirely absent, and almost all surface drainage is captured and drained internally by closed sinkhole depressions. This type of karst is typical of carbonate rocks that

have high intrinsic permeability, such as the Cenozoic limestones in the Atlantic coastal regions, and includes the well-known Floridan aquifer system. In reality, the physical and hydrologic distinctions between fluviokarst and doline karst are not always clearly defined, and many karst terranes have characteristics common to each.

Conduits and Springs

The most distinctive feature of karst aquifers are the typically dendritic or branching networks of conduits that meander among bedding units, join together as tributaries, and increase in size and order in the downstream direction (Palmer, 1991). In the simplest terms, these conduit networks grow by way of a complex hydraulic-and-chemical feedback loop, in which the basic steps are: conduit growth and enlargement → increased hydraulic capacity → increased discharge → enhanced dissolution and physical corrosion → additional conduit enlargement → subsurface piracy of flows in smaller conduits by the larger conduits. In this process, the largest conduits act as master drains that locally alter the hydraulic flow (or equipotential) field so as to capture ground water from the surrounding aquifer matrix, the adjoining fractures, and the smaller nearby conduits (Palmer, 1991, 1999; White and White, 1989) (fig. 2). Depending on their sizes (hydraulic capacity) and organization (interconnection), conduit networks are capable of discharging large volumes of water and sediment rapidly through a karst aquifer (White, 1993). Flow velocities in well-developed and well-integrated conduit networks that range on the order of hundreds to thousands of feet per day are not uncommon (White, 1988).

Karst springs are the natural outlets for water discharging from conduit networks (fig. 3). They typically are developed at a local or regional ground-water discharge boundary—that is, at a location of minimum hydraulic head in the aquifer—often at or near the elevation of a nearby base-level surface stream (White, 1988). The tributary system of conduit drainage typically

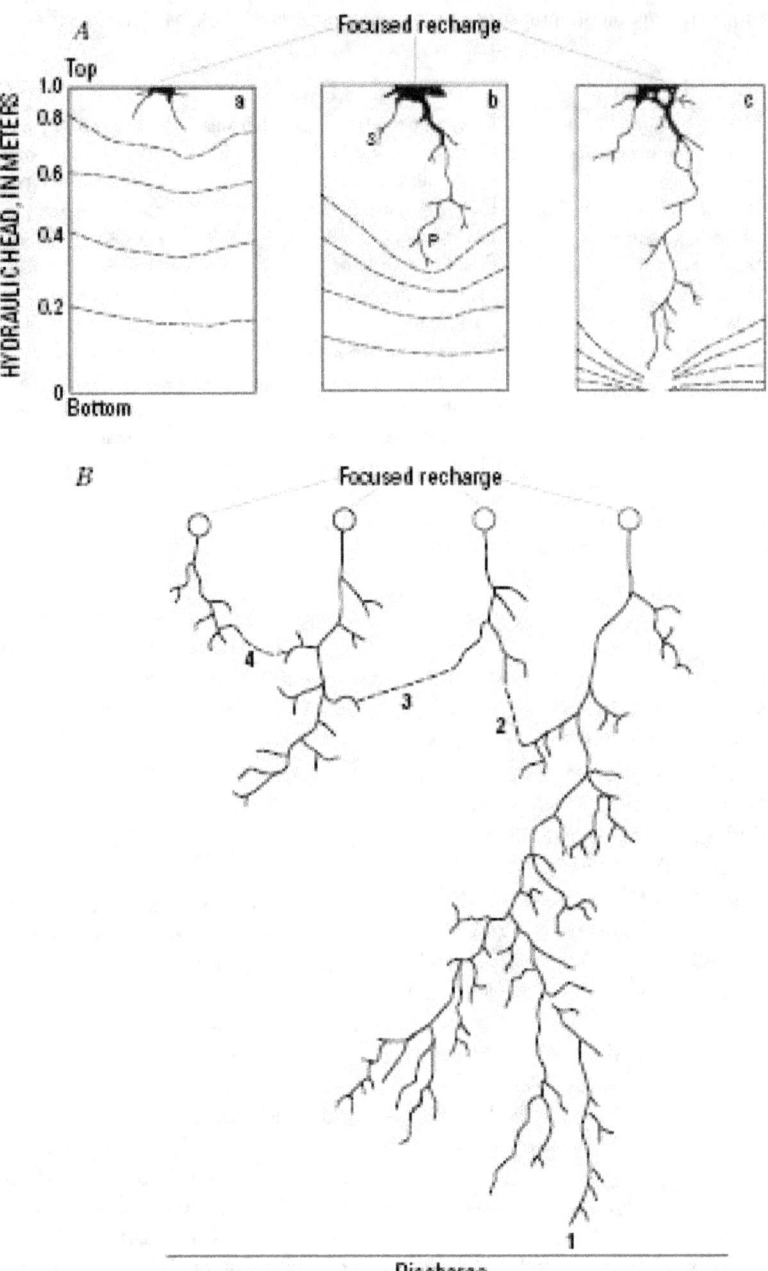

Figure 2. *A*, Diagram showing competitive growth of conduits and distortion of hydraulic flow field: (a) initiation of recharge, (b) change in hydraulic gradient in response to propagation of faster growing primary (P) conduit and slower growing secondary (S) conduit, (c) primary conduit breaks through to discharge boundary, slowing or inhibiting growth of secondary conduit. *B*, Sequence of development of integrated drainage network due to faster growth and breakthrough by primary conduit (1) and subsequent capture of flow and linking of secondary conduits (2–4). (Modified from Ford, 1999, fig. 8.) (Copyright Karst Waters Institute and Dr. Derek Ford, used with permission.)

A

C

B

D

Figure 3.   Photographs showing a variety of physical outlets for karst springs: *A*, Orangeville Rise, southern Indiana; *B*, Whistling Cave Spring, southern Indiana; *C*, Rocky Spring, central Kentucky; *D*, Head-of-Doe-Run Spring, central Kentucky. (Photographs by Charles J. Taylor, U.S. Geological Survey.)

developed in most karst aquifers yields convergent flow to a trunk conduit that discharges through a single large spring (White, 1999). Many karst aquifers, however, have a distributary flow pattern where discharge occurs through multiple spring outlets. This distributary flow pattern generally occurs where there has been enlargement of fractures and smaller conduits located near a stream discharge boundary, where collapse or blockage of an existing trunk conduit or spring has resulted in shifting of flow and development of alternative flow paths and outlets, or where subsurface conduit piracy has rerouted preexisting conduit flow (Quinlan and Ewers, 1989).

Traditionally, springs are classified on the basis of discharge per Meinzer's scale (Meinzer, 1927) and are otherwise characterized on the basis of physical appearances and whether or not the discharge occurs under artesian or gravity flow (open-channel) conditions (U.S. Geological Survey, 2005). From a flow-system perspective, it may be more useful to classify karst springs according to their hydrologic function as outlets

for conduit networks (Worthington, 1991, 1999). In most karst aquifers, one or a few perennial springs, called *underflow springs*, carry the base-flow discharge of conduits (Worthington, 1991). The elevation of the underflow springs exerts much control on the elevation of the water table at the output boundary of the karst aquifer, whereas the matrix hydraulic conductivity and the conduit hydraulic capacity determines the slope of the water table upstream and its fluctuation under differing hydrologic conditions (Ford and Williams, 1989). Other intermittent springs, called *overflow springs*, function as spillover outlets during periods of high discharge. Overflow springs are essentially a temporal form of distributary discharge. As conduits evolve through time and as base levels and water tables are lowered, the upper parts of the karst aquifer may be progressively drained and higher level conduits abandoned (Hess and White, 1989). During high-flow conditions, these higher level conduits may be reactivated and discharge through overflow springs now located at the outlets of former underflow springs.

Karst Recharge

Karst terrane is unique in having multiple sources of recharge that vary considerably in terms of water residence time and in the timing and amounts of water contributed to the conduit network. Sources of karst recharge are categorized as *concentrated* or *diffuse*, and as either *autogenic* or *allogenic* depending, respectively, on whether the recharge originates as precipitation falling on karstic or nonkarstic terrane (Gunn, 1983). These distinctions are important because the relative proportion of concentrated to diffuse recharge generally dictates the distribution and linking together of conduits, and the timing and relative contributions of water fluxes from allogenic and autogenic sources significantly affects the variability in spring discharge and water chemistry (Ford and Williams, 1989).

A cross-sectional diagram of the major sources of recharge that contribute to a typical karst flow system is shown in figure 4. A major source of *concentrated allogenic* recharge to many karst aquifers is water contributed by sinking or losing streams that originate as normal gaining streams in nonkarstic borderlands. A major source of *concentrated autogenic* recharge is surface runoff funneled into sinkhole depressions, which may drain rapidly to the subsurface through throatlike openings called *swallets* or may drain relatively slowly by percolation through a mantle of soil or alluvium. *Diffuse allogenic* recharge may be contributed by interaquifer transfer of water from nonkarstic aquifers, but a more common source is water that drains down the walls of unsaturated (vadose) zone shafts— vertical or near-vertical conduit passages—where karstic rocks are overlain by nonsoluble caprocks such as sandstone (Gunn, 1983).

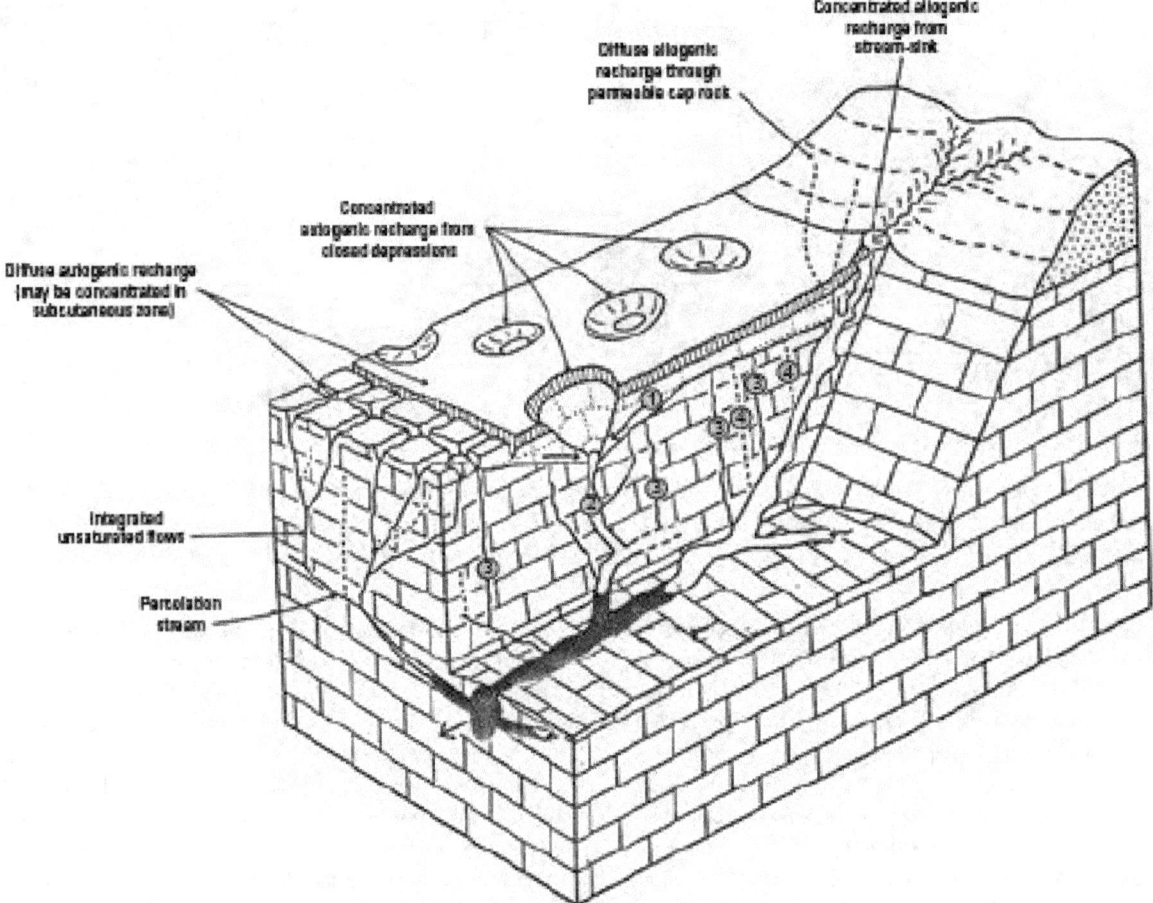

Figure 4. Geologic cross section of a karst basin showing various types of recharge sources: concentrated versus diffuse, and autogenic (recharge that originates as precipitation falling directly on karstic rocks) versus allogenic (recharge that originates as precipitation falling on nonkarstic rocks). Water flows through the unsaturated zone via (1) diffuse flow through soil or unconsolidated surface materials, (2) concentrated flow through solution-enlarged sinkhole drains, (3) diffuse infiltration through vertical fractures, and (4) diffuse infiltration through permeable rock matrix. Subterranean conduits shown as solid black are filled with ground water. (Modified from Gunn, 1986, used with permission.)

In typical studies of karst hydrology, the understandable focus placed on the characterization of concentrated recharge tends to overshadow the fact that most recharge to karst aquifers is contributed by diffuse autogenic recharge—that is, by infiltration through soil—as it is in most other hydrogeologic settings. In a study in Missouri, Aley (1977) estimated that the quantity of water contributed to a karst aquifer from diffuse areal recharge was approximately four times greater than that contributed by all concentrated recharge sources and almost twice that contributed by sinkholes and losing streams combined. Most sinkhole swallets have active inflow only during periods of heavy surface runoff when soil and macropore infiltration capacity is exceeded and, depending on antecedent moisture conditions, the inflow of concentrated recharge by way of swallets may not occur during many storms.

A particularly important source of recharge and storage in most karst aquifers is the *epikarst*—a zone of intensely weathered, fractured, and solution-modified bedrock located near the soil-bedrock contact (Williams, 1983). The thickness and physical hydrogeologic properties of the epikarst are highly variable within and among karst terranes because epikarst development is dependent on stratigraphic variability; bedrock porosity, permeability, and solubility; fracture density; and intensity of weathering. In terms of hydrology, the epikarst functions generally as a leaky perched aquifer zone, providing relatively long-term, diffuse autogenic recharge to conduits (Klimchouk, 2004). Much of the base-flow discharge from karst aquifers to springs and surface streams is water contributed from storage in the epikarst. Chemical hydrograph separation studies have indicated that flushing of water from the epikarst may contribute as much as 50 percent of the water discharging from springs during storms (Trček and Krothe, 2002). Much research has been devoted to the development and hydrologic functioning of the epikarst; however, it remains one of the more poorly understood recharge components of karst aquifers (Aley, 1997; Jones and others, 2004).

Karst Drainage Basins

In typical hydrogeologic studies, a fundamental mapping unit—usually defined by the ground-water basin—is used to characterize the spatial and temporal properties of the aquifer and to construct a conceptual model. For a karst aquifer, the traditional concept of the term "ground-water basin" is somewhat of a misnomer in that it minimizes the highly interconnected nature of surface and subsurface waters and the role of concentrated stormwater runoff as a significant source of recharge. A more appropriate term, and conceptual model, for most karst aquifers is the karst drainage basin (or karst basin)—a mapping unit defined by the total area of surface and subsurface drainage that contributes water to a conduit network and its outlet spring or springs (Quinlan and Ewers, 1989; Ray, 2001). Karst basins differ from conventionally defined ground-water basins—that is, the local ground-water basins described by Toth (1963)—in the following respects:

- Karst basin boundaries do not always coincide with topographic drainage divides, and discharge may or may not always be to the nearest surface stream.

- Recharge near the basin boundaries may flow in a radial or semiradial direction into adjacent basins drained by other underflow springs, and the divides between basins may be indistinct and may shift with changing hydrologic conditions.

- Direct injection of concentrated stormwater runoff and subsurface piracy of surface streamflows constitute a significant portion of the recharge to the basin.

- Most of the active flow is concentrated in the core of the basin, which consists of the conduit network, and is characterized by pipe-full or open-channel hydraulics. Vertical or cascading flow may be significant (Thrailkill, 1985).

- Hydraulic gradients, the number of active conduit flow routes, and directions of ground-water flow may change rapidly with changing hydrologic conditions.

- Directions of ground-water flow do not always conform with the maximum hydraulic gradient inferred by water-level measurements in wells.

- The contributing area and volume of discharged subsurface water changes over time as conduit development, hydraulic capacity, and subsurface piracy increases. In addition, the aquifer carries a substantial sediment load that is constantly changing and can alter flow routes and hydraulic properties of conduits (White, 1988; Dogwiler and Wicks, 2004).

Ray (1999, 2001) proposed that karst basins can be broadly categorized into three functional hydrologic groups on the basis of the hydraulic capacity of their conduit networks and their dominant recharge source (allogenic or autogenic). The three basin groups are defined as:

- Overflow allogenic basins—basins in which the trunk (master) conduit draining the basin is recharged mostly by subsurface piracy of a surface stream(s), but because of limited hydraulic capacity, the surface channel is maintained as a losing stream reach or as an intermittent, storm-overflow route.

- Underflow allogenic basins—basins in which the hydraulic capacity of the trunk conduits has increased to the point that the surface flow is completely diverted underground through streambed swallets, and the surface valley becomes blind.

- Local autogenic basins—basins in which all surface flow has been captured by subsurface piracy, and the trunk conduit is recharged almost exclusively by infiltration through the soil and internal sinkhole drainage.

These basin categorizations apply best to shallow, unconfined karst aquifers in fluviokarst settings, but also describe basins in doline karst and in deeper partly confined karst aquifers such as the Madison Limestone aquifer (Greene, 1997) in South Dakota, which is characterized by overflow allogenic basins recharged by sinking streams draining a structurally uplifted recharge area. A progressive sequence of karst basin development—from overflow allogenic to underflow allogenic to local autogenic—may occur in many karst terranes over geologic time as karstification and subsurface piracy of surface streams increases (Smart, 1988; Ray, 2001).

Hydrogeologic Characterization

As in other complex hydrogeologic settings, a proper hydrogeologic characterization of karst drainage basins is the key to understanding and estimating water fluxes. As applied within the framework of a karst conceptual model, this requires the acquisition of data needed to characterize the extent and overall effects of conduit-dominated flow, multiple discrete inputs and outputs for water, and spatial and temporal variability in recharge, storage, and flow. Water-tracing tests, typically done using fluorescent dyes, are the most effective means of determining subsurface conduit connections between karst drainage features such as sinkholes and springs, directions of ground-water flow in the karst aquifer, boundaries of karst ground-water basins, and the hydraulic properties of conduits (Mull and others, 1988; White, 1993). The analysis of spring discharge hydrographs and temporal variations in the chemical or isotopic composition of spring water provide data needed to characterize the recharge, and storage and discharge functions occurring in karst aquifers and to provide additional insights into the structure of conduits at basin-to-regional field scales (Ford and Williams, 1989; White, 1993).

Other, more conventional hydrogeologic data-collection methods—including those described in other chapters of this report—also may be used in the study of karst aquifers if these methods are applied within the framework of a karst conceptual model. Careful consideration must be given to the field scale of collected hydrologic measurements and to whether the measurements obtained by use of a particular method are representative of the conduit-dominated flow components of the aquifer, the aquifer matrix or nonconduit flow component, or a composite of both.

In addition to the topographic, structural, and stratigraphic characteristics that are necessary to define the physical hydrogeologic framework, White (1999) proposes six basic hydrologic properties needed for the evaluation of karst basins: (1) the area of the karst basin, (2) allogenic recharge, (3) conduit carrying capacity, (4) matrix and fracture system hydraulic conductivity, (5) conduit system response, and (6) conduit/fracture coupling. A water budget is suggested here as a seventh additional characteristic for evaluation. Information collected about each of these seven karst basin features will contribute to the identification and estimation of fluxes between surface water and ground water in karst terranes.

Area of the Karst Drainage Basin

Various methods have been used to estimate the recharge or contributing areas of karst springs (Ginsberg and Palmer, 2002), but dye-tracer tests provide the most effective means of identifying the point-to-point connections between flow inputs (sinkholes or sinking streams) and outputs (springs) needed to actually define the boundaries of karst drainage basins (White, 1993; Ray, 2001). Dye-tracer tests can be done at multiple input sites by injecting different fluorescent dyes either simultaneously or sequentially. As tracer-inferred ground-water flow directions are determined and the number and distribution of tracer-determined flow paths increase, the boundaries, approximate size, and shape of the basin under study can be delineated with increasing levels of confidence. To fully delineate the boundaries of the area contributing recharge to a particular spring, dye-tracer tests need to be planned and conducted in strategic locations so that the results obtained "push" the point-to-point connections established between the spring and its contributing inputs (for example, sinkholes) toward the anticipated locations of subsurface drainage divides. The presence of these drainage divides are inferred where the trajectories of plotted dye-tracer flow paths indicate a divergence in subsurface flow directions, that is, identify areas where subsurface flows are being routed to springs draining other adjacent karst basins. The geographic distribution of these inferred subsurface drainage divides constrains the boundaries of the karst basin under study (fig. 5).

Tracer-inferred flow paths can be plotted as straight lines between input and resurgence sites, or preferably, as curvilinear vectors that depict a tributary drainage system more visually representative of the natural conduit network (Ray, 2001). Other hydrogeologic mapping data such as cave surveys or contoured water-level maps can be used as an aid in the planning and interpretation of dye-tracer tests; for example, the locations of major ground-water conduits often are correlated with the positions of apparent troughs in the potentiometric surface or water table, which are thought to represent a locus of maximum ground-water flow (Quinlan and Ewers, 1989). Karst mapping studies that illustrate various applications of these techniques include those of Crawford (1987), Mull and others (1987), Vandike (1992), Bayless and others (1994), Schindel and others (1995), Imes and others (1996), Jones (1997), Taylor and McCombs (1998), and Currens and Ray (1999).

Dye-tracer tests have routinely shown that conduit flow paths commonly extend beneath topographic drainage divides and, in some places, beneath perennial streams, and that surface runoff draining into sinkholes or sinking streams in one topographic basin (watershed) may be transferred via subsurface flow routes into adjacent topographic basins (Ray, 2001) (fig. 6). In karst terranes, mapping of the contributing areas of springs and surface streams, identification and estimation of water fluxes and, in particular, estimation of water budgets for either surface or subsurface drainage basins, are critically dependent on identifying and delineating the areas that indicate this "misbehaved drainage" (White and Schmidt, 1966; Ray, 2001). Dye-tracer tests are the most reliable method of obtaining this information. For example, dye-tracer tests were used to conclusively demonstrate that the USGS Hydrologic Unit (watershed) boundaries delineated for the Barren River basin in central Kentucky using topographic drainage divides encompass approximately 220 square kilometers (85 square miles) of surface drainage that actually contributes water to the adjacent Green River basin via subsurface conduits (Ray, 2001) (fig. 6).

Allogenic Recharge and Conduit Carrying Capacity

As previously noted, a significant component of recharge to underflow and overflow allogenic karst basins is the water contributed by subsurface piracy of surface streams, and it is this concentrated allogenic recharge that largely influences the discharge and water-chemistry changes indicated by karst springs during and after storms. Quantifying the allogenic recharge subbasin area and the sum of the inputs from individual sinking or losing streams defines an important characteristic of the hydrology of a karst basin. Geographic Information System (GIS) technology provides a convenient way of delineating the catchment areas of all sinking or losing streams that contribute to a karst basin and of estimating the relative proportion of allogenic recharge subbasin area to autogenic recharge (sinkhole-dominated) subbasin area (Taylor and others, 2005). In theory, all of the allogenic recharge contributed to a karst basin can be measured by synoptic gaging of discharge in the stream channels directly above the locations of terminal swallow holes. When evaluated with discharge measurements from the basin's outlet springs, the measured allogenic inputs provided by each sinking or losing stream can be used to evaluate conduit-carrying capacity (White, 1999) in the following manner:

In underflow allogenic basins, the hydraulic capacities of the conduits are defined by the following relation:

$$Q_c > Q_{a(max)} \qquad (1)$$

where

$\quad Q_c \quad$ is the carrying capacity of the conduits

and

$\quad Q_{a(max)} \quad$ is the maximum discharge of the surface stream(s) contributing recharge to the conduits.

In this particular instance, the carrying capacity of the conduit network always exceeds the maximum input contributed by the allogenic stream recharge, and surface flows are completely diverted underground by one or more swallow holes shortly after crossing onto karstic bedrock. This case describes a classic sinking stream.

In overflow allogenic basins, the carrying capacities of the conduits are defined by one of two relations (eqs. 2 or 3):

$$Q_{a(base)} > Q_c, \qquad (2)$$

where

$\quad Q_{a(base)} \quad$ is the base-flow discharge of the allogenic surface stream.

In this case, the carrying capacity of the conduits cannot accommodate the base-flow discharge of the allogenic stream, and perennial surface flow occurs in the channel despite flow losses through streambed swallow holes. This case describes a classic losing stream.

$$Q_{a(max)} > Q_c > Q_{a(base)}. \qquad (3)$$

In this case, the carrying capacity of the conduits can accommodate all of the base-flow discharge from the allogenic stream, but stormflow discharge often exceeds the capacity of the conduits, overtops swallow holes, and results in continuation of flow down the channel. This case describes an intermittent sinking stream, often characterized as a "dry-bed stream" (Brahana and Hollyday, 1988). The reactivation of swallow holes as sink points often occurs in a successive manner as surface flow overtops upstream swallow holes first and reaches or overtops the farthest downstream swallow holes only during the largest storms (George, 1989).

White (1999) makes the interesting suggestion that determining the critical flow threshold when $Q_a = Q_c$ would be a meaningful way of characterizing conduit permeability; however, it would require gaging the discharge in sinking streams above the terminal swallow holes at the exact time that the swallow holes are filled and overtopped. There are practical difficulties involved in obtaining such measurements, not only with regard to the timing of the measurements, but because flow in the channels of many sinking streams often is lost progressively through a series of swallow holes; for example, the Lost River basin of southern Indiana (Bayless and others, 1994), or because clogging of the swallow holes with sediment or debris is a factor that controls the rate of inflow (Currens and Graham, 1993).

Matrix and Fracture System Hydraulic Conductivity

Because of combined permeability provided by matrix, fracture, and conduit-flow components, the timing and amount of response to hydraulic stresses varies greatly from place to place within a karst aquifer. Investigation of the hydraulic conductivity of the matrix and fracture components is typically performed with conventional hydrogeologic tools. Matrix permeability can be determined using laboratory permeability tests done on representative rock core samples. Fracture hydraulic conductivity is best determined using

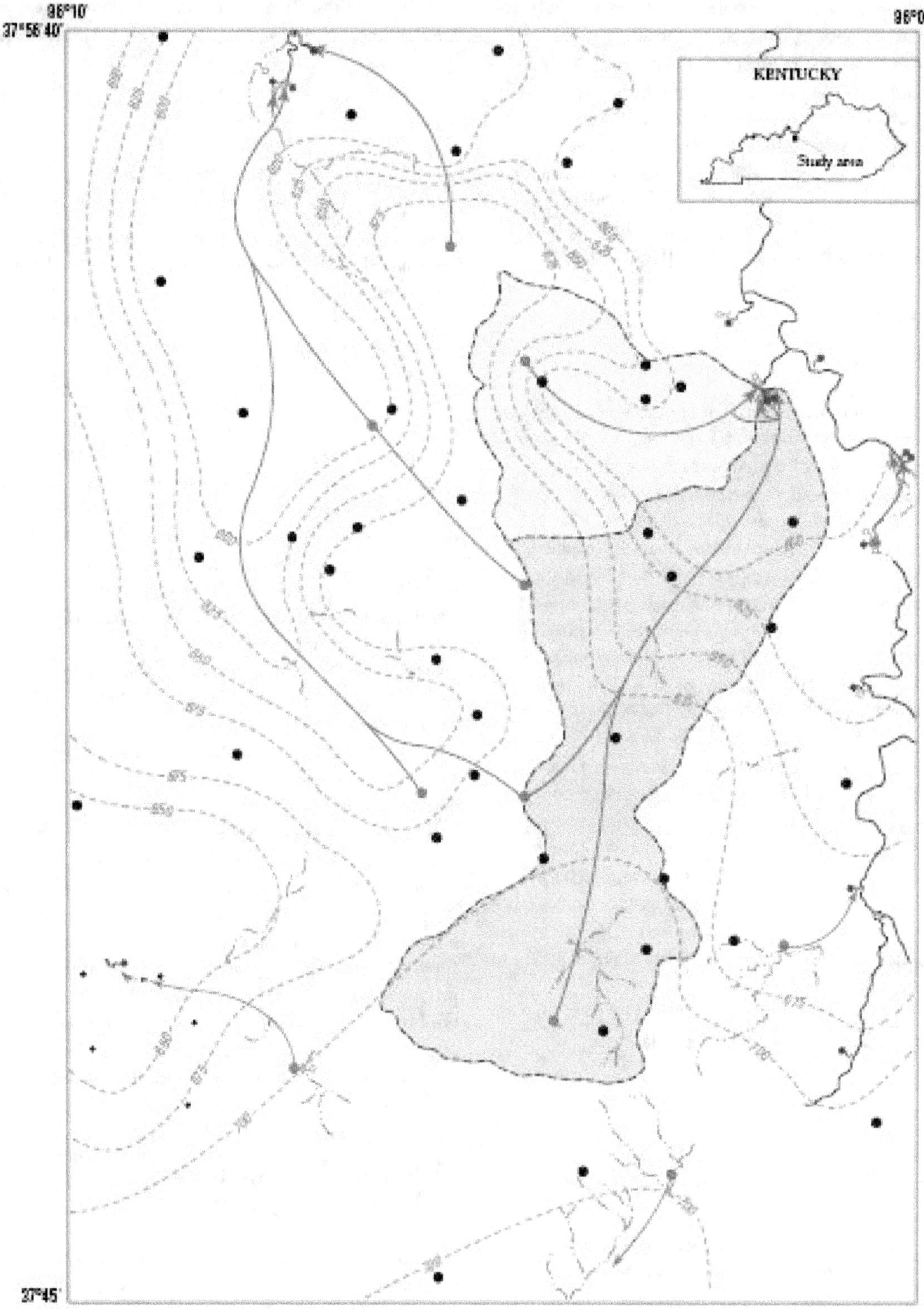

Base from U.S. Geological Survey Digital Data, 1:24,000, 1986
Mercator Projection

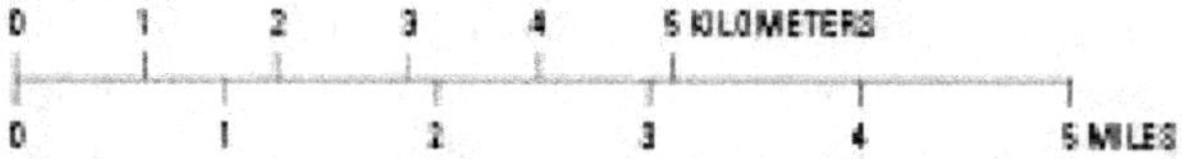

EXPLANATION

☐ Northern ground-water basin

☐ Southern ground-water basin

- - 600 - - Inferred potentiometric-surface contour—Shows altitude at which water level is expected to stand in tightly cased wells completed exclusively in the St. Louis Limestone. Contour interval 25 feet. Datum is North American Vertical Datum of 1988 (NAVD 88).

- - - - - - Ground-water basin boundary—Appropriate location of ground-water divide defined by topographic, geologic, and hydrologic features that influence the direction of ground-water flow.

- - - - - - Dye flow path—Shows inferred route of dye tracer in karst aquifer and confirmed hydraulic connection between dye-injection site and dye-recovery site. Dashed line indicates intermittent flow route to an overflow spring. Number indicates dye-tracing test.

- - - • Intermittent stream and terminal sink point (swallow hole)

• Well

⊛ Dye-injection site

➤ Dye-recovery site

⌢● Perennial (underflow) spring

⌢○ Intermittent (overflow) spring

◦-• Karst-window—Perennial spring and sinking stream

Figure 5 (above and facing page). Part of map showing dye-tracing flow paths (red curvilinear vectors) used to constrain the boundaries for two karst spring subbasins (orange, yellow shading). Dashed blue lines are water-table contour lines, which provide additional information useful in mapping the basin boundaries and interpreting subsurface flow paths (modified from Taylor and McCombs, 1998).

straddle-packer hydraulic tests and borehole flow meters (Sauter, 1991). Conventional aquifer tests (time-drawdown, distance-drawdown, or slug tests) provide a measurement of the integrated local matrix and fracture system transmissivity. Borehole geophysical methods, including cross-borehole tests, also provide valuable data to assist with permeability and flow characterization at local to subbasin scales (Paillet, 2001).

Analysis of karst aquifer test data using conventional Darcian analytical methods may provide erroneous results, and special consideration should be given to the possible effects of slow-flow and quick-flow karst components on the hydraulic responses represented by the well-hydraulic test data. Streltsova (1988) reviews aquifer-test methods best suited to investigations of heterogeneous aquifers such as karst. If the test well penetrates large solutional openings or conduits, the hydraulic conductivity (or transmissivity) and storage coefficients of these should be evaluated separately from those of fractures (Greene and others, 1999).

Comparative studies of hydraulic properties measured in different karst aquifers have shown that, regardless of the range of porosity measured in the aquifer matrix, conduits

typically account for less than 1 percent of the porosity of the aquifer, but more than 95 percent of the permeability (table 2) (Worthington and others, 2000). As in studies of many fractured rock aquifers, there is a general tendency for measured hydraulic conductivities to increase with increasing field scale (Sauter, 1991). Typically, the distribution of hydraulic conductivity and other properties is related to lithostratigraphic facies changes or other physical changes in the characteristics of the bedrock matrix (Rovey and Cherkauer, 1994).

Conduit System Response

Conduit system response may be evaluated using: (1) quantitative water-tracing tests to determine traveltime and tracer-breakthrough characteristics, (2) recession analysis of spring discharge hydrographs (White, 1999), (3) evaluation of the ratio between peak storm discharge and base-flow discharge (Q_{max}/Q_{base}) of karst springs, (4) chemical hydrograph separation, and (5) hydrologic pulse analysis—analysis of changes in spring discharge and water-quality constituents in response to storms (Ryan and Meiman, 1996; Katz and others, 1997).

Early studies of the variation in spring discharge and water chemistry led to the suggestion that karst springs and aquifers could be categorized along a hydrologic continuum defined by conduit-dominated and diffuse-dominated end members (Shuster and White, 1971; Atkinson, 1977; Scanlon and Thrailkill, 1987) with the observed hydrologic response differing according to the proportion of conduit-to-nonconduit permeability (fig. 7). So-called conduit-dominated karst springs typically exhibit rapid changes in discharge and wide-ranging changes in water chemistry in response to precipitation input (fig. 8). In contrast, so-called diffuse-dominated karst springs respond more slowly to precipitation input and exhibit more buffered, gradual changes in discharge and water chemistry. These distinctions seem to be applicable in a broadly descriptive context and are still used as a convenient way of characterizing karst flow systems.

More recent studies have indicated that karst spring discharge and water chemistry responses are influenced by temporal variability in the proportion of recharge contributed from diffuse and concentrated sources (White, 1999), and by the timing and volume of water contributed from conduit, fracture, and matrix flow components that reflect the range of transmissivities present in the karst basin or aquifer (Doctor and Alexander, 2005). Many karst springs and aquifers are observed to exhibit a dual or triple hydrologic response to precipitation defined by: (1) an initial rapid flow response created by water transmission in conduits greater than 5 to 10 millimeters in diameter where velocities generally exceed 0.001 meter per second, followed by (2) a secondary, slower flow response created by water transmission in intergranular pore spaces, smaller aperture fractures, and solutional openings within the aquifer matrix where velocities are less than 0.001 meter per second (Worthington, Davies, and Ford, 2000), and (3) a transitional response period between these

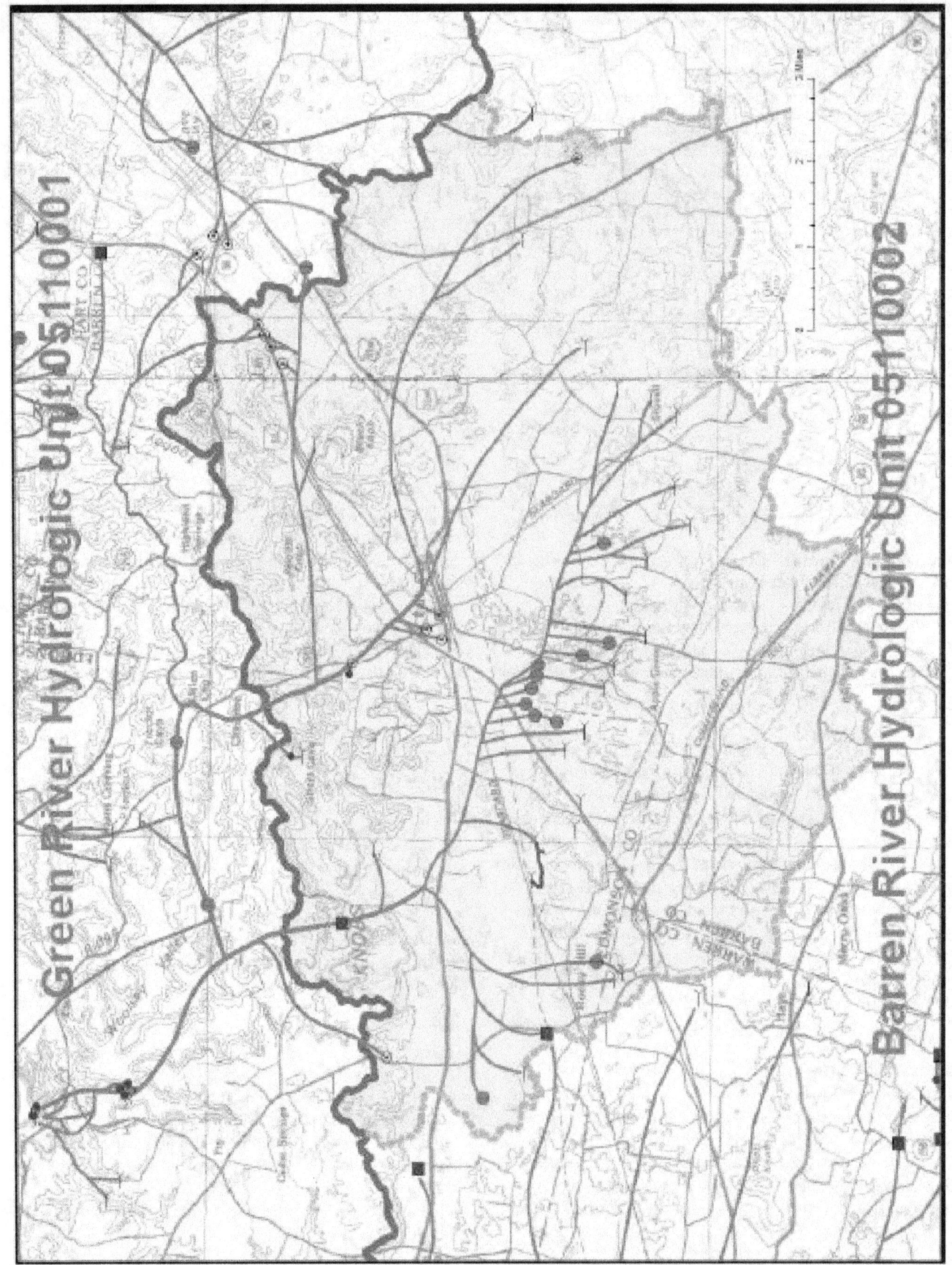

EXPLANATION

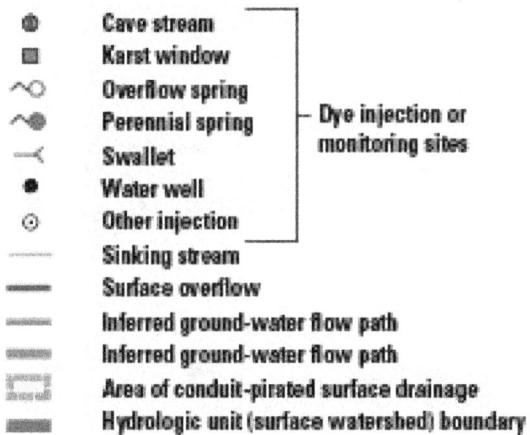

⊕	Cave stream
▣	Karst window
⌒○	Overflow spring
⌒◉	Perennial spring
⤙	Swallet
●	Water well
⊙	Other injection

Dye injection or monitoring sites

⸺⸺	Sinking stream
▬▬	Surface overflow
▬▬	Inferred ground-water flow path
▬▬	Inferred ground-water flow path
▥▥	Area of conduit-pirated surface drainage
▬▬	Hydrologic unit (surface watershed) boundary

Figure 6 (above and facing page). Subsurface conduit piracy of surface drainage from part of the Barren River watershed to springs discharging to the Green River watershed. (Courtesy of Joe Ray, Kentucky Division of Water, map modified from Ray and Currens, 1998.)

two. Accordingly, the alternative terms "quick flow," "slow flow," and "mixed or intermediate flow" now are used often to describe the range of hydrologic responses exhibited by a karst spring or aquifer (White, 1993). Various methods of spring hydrograph analysis, summarized later in this chapter, may be applied to investigate and quantify these changes in karst hydrologic responses.

One simple method of quantifying and evaluating the "flashiness" of the conduit system response is to determine the ratio of maximum peak-flow to base-flow spring discharge (Q_{max}/Q_{base}): it is a function of storm intensity and conduit organization or interconnectivity (White, 1993). Springs dominated by a quick-flow response typically exhibit Q_{max}/Q_{base} ratios in the range of 40 to 100, whereas ratios of about 1 to 3 and 7 to 10, respectively, are exhibited by springs dominated by a slow flow response and by intermediate or mixed flow response (White, 1993). The timing of these changes in hydrologic response depends on the size of a karst basin, the distances between flow inputs and outputs, and on the internal organization of its conduit network (White, 1993). The response time, t, determined by fitting an exponential function to the recession limb of the spring hydrograph, also seems to indicate a wide range in values that cluster into distinctive groups characteristic of each hydrologic response type.

Conduit/Fracture Coupling

Under normal base-flow conditions, conduits act as low-hydraulic resistance drains that locally alter the hydraulic flow (or equipotential) field so as to capture ground water from the surrounding aquifer matrix and adjoining fractures (White, 1999). The flux of water between conduit-flow and nonconduit-flow components is a complex head-dependent process and may be reversible when conduits fill completely and pressurize under certain storm-flow conditions. Water flux reversal also can be induced by backflooding of surface streams, wherein surface water enters conduit passages by way of underflow and overflow springs and results in hydraulic damming. In either instance, the injection of water from the conduits back into the aquifer matrix constitutes an unusual type of aquifer recharge and bank storage, which has been well documented, for example, in the Green River-Mammoth Cave karst aquifer system in Kentucky (Quinlan and Ewers, 1989). As stormwater or flood pulses are drained rapidly through the conduits, spring discharge returns to base-flow conditions, and the normal flux resumes as the dominant source of recharge shifts to water contributed from longer term storage in the epikarst, bedrock matrix, fractures, and smaller tributary conduits.

The effectiveness of the coupling between conduit and fracture components, combined with the hydraulic conductivity of the matrix/fracture system, control the rate of movement of water into and out of storage after storms or floods and during base-flow conditions (White, 1999). The conduit/fracture coupling can be evaluated by: (1) deconvolution of spring hydrographs, (2) comparisons of storm-related hydrograph response in springs or observation wells in the manner described by Shevenell (1996), and (3) evaluation of unit base flow.

The unit base flow (UBF), or base-flow discharge per unit area, is a particularly useful measurement derived from the concept that surface-stream watersheds of similar size (area) located in similar hydrogeologic settings and climates will generate approximately equal quantities of base-flow runoff (Quinlan and Ray, 1995). Applied to karst basins, the UBF represents the amount of water discharged from long-term ground-water storage, as controlled by the coupling between the conduits and the diffuse-flow component. Its value is best calculated by using dry-season, base-flow spring discharge measurements (Quinlan and Ray, 1995). Table 3 lists the range of UBF values calculated for several spring basins in Kentucky. UBF values are useful in estimating the basin areas of springs in similar hydrogeologic settings whose basin boundaries are unknown or untraced, and help identify anomalous recharge or storage characteristics for a spring basin under study (White, 1993; Quinlan and Ray, 1995).

Table 2. Comparison of porosity and permeability measurements in various karst aquifers (after Worthington, 1999).

[%, percent; m/s, meter per second]

Karst area	Porosity (%)			Hydraulic conductivity (m/s)	
	Matrix	Fracture	Conduit	Matrix	Fracture
Smithville, Ontario	6.6	0.02	0.003	1×10^{-10}	1×10^{-5}
Mammoth Cave, Kentucky	2.4	0.03	0.06	2×10^{-11}	1×10^{-5}
Devonian Chalk, England	30	0.01	0.02	1×10^{-6}	4×10^{-4}
Nohoch Nah Chich, Yucatan, Mexico	17	0.1	0.5	7×10^{-5}	1×10^{-1}

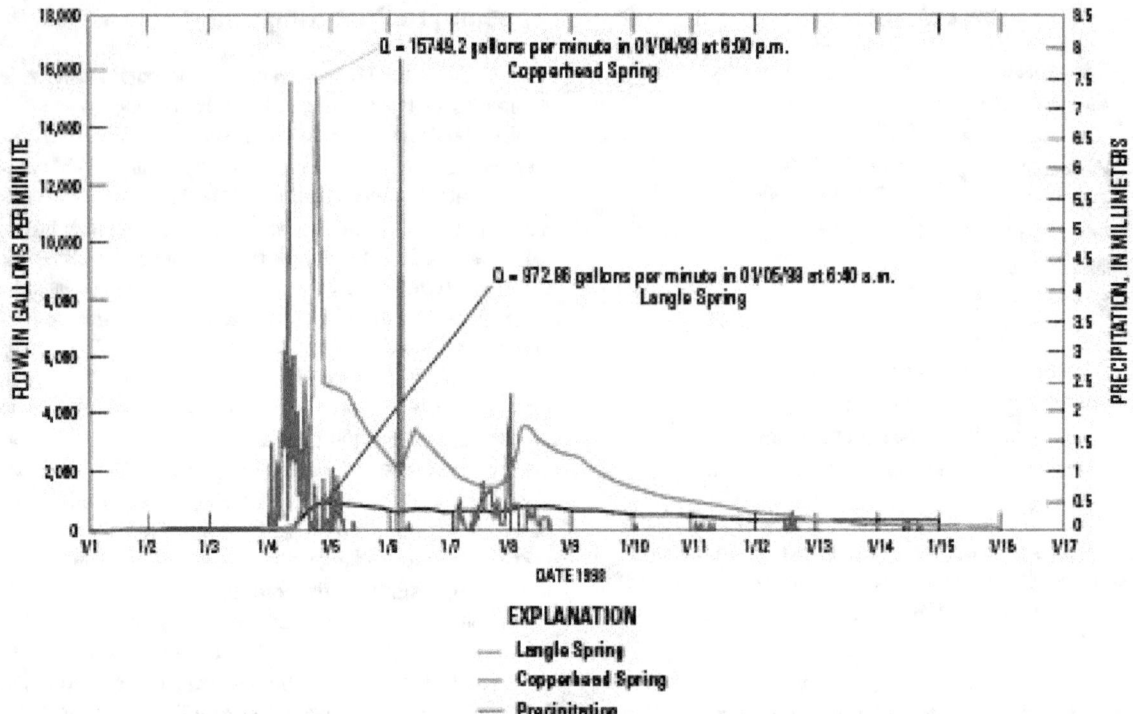

Figure 7. Variable response of springs to precipitation. Copperhead Spring hydrograph shows rapid conduit-dominated flow response. Langle Spring hydrograph shows slow diffuse-dominated flow response. These are related to the relative proportion of conduit permeability to nonconduit permeability (courtesy of Van Brahana, University of Arkansas).

Water Budget

Water budgets typically are written with the instantaneous flows integrated over a specified period of time, which can be a water year (season to season), a season, the duration of a single storm, or any other period (White, 1988). Published examples of water-budget calculations for karst aquifers include Bassett's (1976) study of the Orangeville Rise spring basin in south-central Indiana, and Hess and White's (1989) study of the spring-fed Green River within the boundaries of Mammoth Cave National Park in Kentucky. Other examples are discussed by Milanović (1981a, b) and Padilla and others (1994). Equations may take a variety of forms depending on the purpose and the hydrologic terms that can be estimated or must be evaluated. A simple, conventional water-budget equation for a karst basin or aquifer may be written in the form:

$$I = O + ET + \Delta S, \qquad (4)$$

where

I is precipitation,

O is basin or spring discharge,

ET is evapotranspiration,

and

ΔS is change in ground-water storage (Bassett, 1976).

Using this equation, a water balance can be obtained by summing the values for O, ET, and ΔS and subtracting the resulting value from I. The results are expressed as the percentage of rainfall unaccounted for (positive I values), or in excess of the balance (negative I values) (table 4).

A water-budget equation also can be written to express the change in storage occurring as a result of a storm:

$$Q_i - Q_o = \pm \Delta V / \Delta t, \qquad (5)$$

where

Q_i is the total inflow or recharge contributed by the storm,

Q_o is the outflow discharge,

ΔV is the change in storage,

and

Δt is the time period of the storm (Ford and Williams, 1989).

Antecedent precipitation and soil-moisture conditions are influential in determining the magnitude of Q_i and Q_o.

More complex water-budget equations can be developed to include additional karst hydrologic factors. White (1988), for example, describes how a water budget developed for an allogenic overflow karst basin might include terms for the input by sinking streams (the allogenic recharge), internal runoff (sinkhole drainage), diffuse infiltration (through soil, epikarst, and bedrock matrix), and positive or negative changes in ground-water storage. In these types of calculations, allogenic

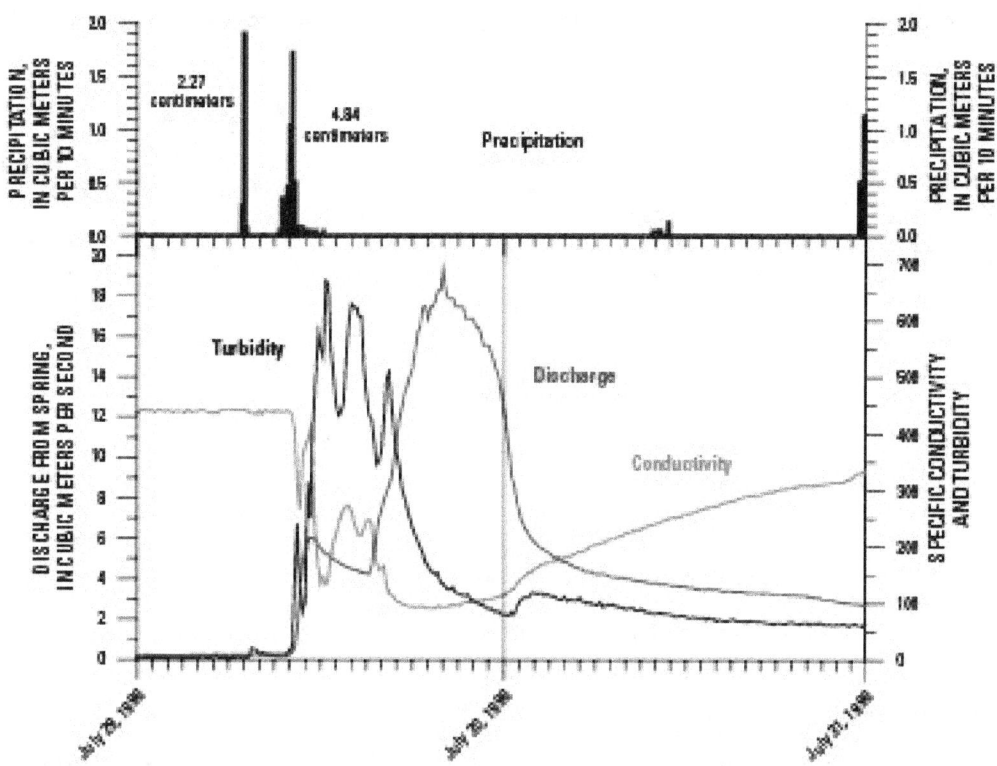

Figure 8. Water-quality changes in a karst spring related to allogenic recharge: Precipitation-stormwater runoff causes a rise in turbidity and a decrease in specific conductance prior to and during peak spring discharge. After passage of the stormwater recharge pulse, conductivity increases as the spring discharge returns to base flow (discharge of water contributed from storage by the slow diffuse-flow karst component). (Courtesy of James Currens, Kentucky Geological Survey.)

recharge from sinking or losing streams can be directly measured, at least in theory, and estimation of the contribution of autogenic (sinkhole) recharge is more problematic. Change in storage typically is estimated from analysis of spring recession hydrographs using the methods described by Kresic (1997), or estimated in terms of net head change in the aquifer on the basis of water-level measurements from observation wells.

Spring-Discharge Hydrograph Analysis

Spring discharge represents an integration of the various processes that govern recharge, storage, and throughflow in a karst basin upstream from its outlet (Kresic, 1997). Analysis of the spring-discharge hydrograph makes it possible to obtain valuable insights into hydraulic stresses acting on the basin, to evaluate basin flow characteristics, and to estimate average basin hydraulic properties (Bonacci, 1993; Baedke and Krothe, 2001; Pinault and others, 2001). A wide variety of graphical, time-series, and spectral analysis techniques have been applied that are beyond the scope of discussion of this chapter. Many of these techniques are reviewed by White (1988) and Ford and Williams (1989).

Analysis of the recession period of the spring discharge hydrograph is one of the simpler and more useful methods to apply to karst studies because it provides information about the volume of water drained from the karst basin over time after peak flows and changes in the rate of discharge that may indicate thresholds and limits in aquifer flow regimes (Doctor and Alexander, 2005). A step-by-step review of the recession analysis technique is presented by Kresic (1997). Its application to the determination of karst basin flow and hydraulic characteristics is summarized here.

Basin Flow Characteristics

Interpretation and analysis of a spring hydrograph assumes that: (1) the discharge of the spring is controlled by input events such as a high-intensity precipitation event or a recharge event at a sinking stream, and (2) the shape of the hydrograph is controlled by flow through various pathways that have different conductivities and velocities (Milanović, 1981b). By using recession analysis, it is possible to identify whether the overall basin flow characteristics are dominated by quick flow (conduit-dominated flow), slow flow

Table 3. Range of Unit Base Flow (UBF) values (designated here as normalized base flow or NBF) in spring basins in Kentucky (modified from Quinlan and Ray, 1995).

[cfs, cubic feet per second]

Autogenic recharge group	Spring name	Spring identification number	Basin area (miles²)	Summer base flow (cfs)	Normalized base flow, NBF (cfs/mile²)	Geometric mean NBF (cfs/mile²)
1. With up to 25 percent allogenic recharge from sandstone-capped ridgetops or near-surface, leaky, chert aquitard	Lavler Blue Hole	21	10.2	2.1	0.21	
	Garvin-Beaver	22	7.2	1.7	0.24	
	Echo River	--	8.8	1.8	0.21	0.20
	Lost River	--	55.2	12.0	0.22	(0.21)
	Pleasant Grove	--	16.1	2.5	0.16	
	Shakertown	--	19.0	3.6	0.19	
2. With significant allogenic recharge from carbonate terrane	Gorin Mill	23	152	25.1	0.17	
	Turnhole	3	90.4	14.3	0.16	0.17
	Graham	21	122	20.8	0.17	
3. With locally thick, areally significant sand and gravel cover	Rio	13	5.2	4.7	0.91	
	Rio	13	6.5	4.7	0.72	
	McCoy Blue	1	36.1	12.3	0.34	
	Roaring	6	10.8	11.9±1	1.19	0.70
	Roaring				1.01	(0.75)
	Johnson	7	17.5	11.0	0.63	
	Jones School	19	3.9	2.3	0.59	
	Jones School	19	2.9	2.3	0.79	
4. With much interbedded shale, Bluegrass Region	Royal	--	25.0	2.8	0.11	
	Russell Cave	--	6.4	1.0	0.15	0.11
	Garretts	--	7.4	0.5	0.07	

(diffuse-dominated flow), or mixed flow, and to evaluate the timing and magnitude of changes in spring discharge that correspond to changes between these flow regimes (fig. 9).

Analysis of a spring discharge hydrograph to determine the flow regimes of the karst basin is done through methods presented by Rorabaugh (1964) and Milanović (1981a,b). Even though these methods are based on Darcian theory, the hydrograph analysis methods have been successfully applied to many studies of karst basins (Baedke and Krothe, 2001; Shevenell, 1996; Padilla and others, 1994, Sauter, 1992; and Milanović, 1981a). The method of characterizing karst flow regimes is based on the equation below, whereby the recession curves of spring hydrographs are analyzed to calculate the value of α, the recession slope:

$$Q_t = Q_o \, e^{-\alpha(t-t_o)} \qquad (6)$$

where

t is any time since the beginning of the recession for which discharge is calculated,

t_o is the time at the beginning of the recession, usually set equal to zero,

Q_t is spring discharge at time t,

Q_o is spring discharge at the start of the recession (t_o),

and

α defines the slope, or recession constant, that expresses both the storage and transmissivity properties of the aquifer.

By using a semilog plot of discharge and time during a spring's recession curve, one can easily determine a characteristic α value that defines the recession curve slope. For some hydrographs, one α value may be obtained that is sufficient to describe the slope of the recession curve. It is not uncommon, however, for karst springs to exhibit two to three major changes in slope on a single hydrograph recession limb, and here it is advantageous to evaluate each slope change and its corresponding α value. A common interpretation of these changes is that the first and steepest slope represents the transmission of the

Table 4. Water budget calculations for Orangeville Rise spring, southern Indiana (modified from Bassett, 1976). Used with permission of the National Speleological Society (www.caves.org).

Interval	Duration (days)	I*	O*	O/I (%)	ET*	ΔS*	Bal.** (%)
June 1 August 20, 1972	81	26.4	3.8	14	26.4	-2.8	-3.8
August 21 October 29	70	20.7	2.9	14	20.7	+1.4	-21
October 30 December 5	37	15.3	5.5	36	3.2	+2.3	28
December 6 January 18	44	11.4	10.8	95	2.8	+0.1	-18
January 19 March 2	43	6.6	7.9	120	0.8	-0.4	-26
March 3 May 6	65	32.4	19.7	61	3.7	+1.3	24
May 7 July 5, 1973	61	20.1	10.0	50	7.2	-1.6	6

*Units are acre-feet × 10^3.

**Bal. = I − (O + ET + ΔS), expressed as a percentage of I.

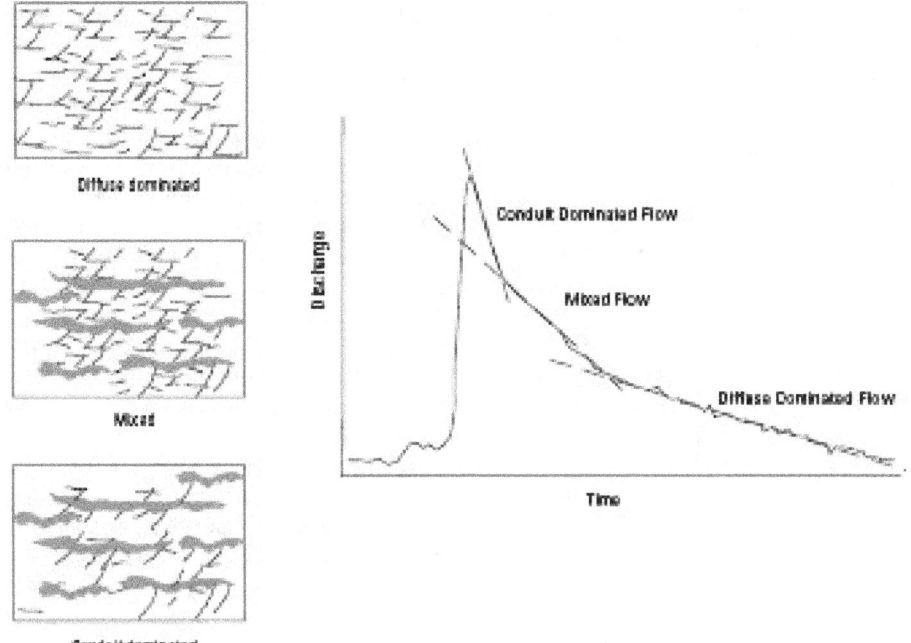

Figure 9. Conceptual spring hydrograph showing changes in slope and dominant flow regime (conduit, mixed, diffuse) due to differing hydraulic responses (artwork by Earl Greene, U.S. Geological Survey).

main stormwater runoff pulse through the largest conduits. This often is followed by a change to a less steep, intermediate recession slope interpreted as marking the beginning of depletion of the stormwater pulse and (or) the spring discharge being composed of a mixture of stormwater and stored ground water discharging from smaller conduits and larger fractures. The final change in slope on the recession curve signals the return to base-flow conditions wherein the spring discharge is composed of ground-water stores discharging from a network of smaller fractures and bedrock matrix.

As noted previously, these differences in spring flow characteristics sometimes are referred to as the quick-flow (or conduit-dominated) response, the intermediate flow response, and the slow-flow (or diffuse-dominated) response. The α value calculated for a spring discharge recession curve, or for each "slice" of a multisloped recession curve, typically takes on a characteristic value or range indicative of each type of flow regime (table 5). For example, all three karst flow regimes (quick-flow or conduit-dominated-flow, mixed flow, and slow-flow or diffuse-dominated flow) are evident in the discharge hydrograph for San Marcos Springs from the Edwards aquifer in Texas (fig. 10).

Basin Hydraulic Properties

Bonacci (1993) and Baedke and Kroethe (2001) have suggested that it is possible to estimate the average transmissivity of the karst basin by using spring-discharge hydrograph analysis, again following the methods of Rorabaugh (1964) and Milanović (1981a), by applying the equation

$$\frac{T}{S_y} = \frac{\log\left[\dfrac{Q_1}{Q_2}\right]}{(t_1 - t_2)} \frac{L^2}{1.071} \qquad (7)$$

where

T	is aquifer transmissivity,
S_y	is specific yield, Q is discharge,
t	is time,

and

L	is the effective karst basin length.

Results obtained from aquifer (well hydraulic) test analysis may be used to estimate the storage (S_y) parameter, and Shevenell (1996) and Teutsch (1992) measured the linear distance from the karst spring to the farthest basin drainage divide to obtain a value for L. The transmissivity estimate obtained using this method needs to be compared to values determined from aquifer tests and quantitative dye-tracer tests.

Chemical Hydrograph Separation

Analysis of the flux of dissolved ionic species or isotopes in spring discharge during storms provides a useful means of identifying water fluxes contributed by different sources of recharge and quantifying their proportions in spring discharge. Although a variety of naturally occurring isotope and geochemical tracers may be used (Katz and others, 1997; Katz, 2005), the method requires that there be a distinctive difference in isotope or geochemical composition between water discharged at base flow and that discharged during storm-pulse

Table 5. Characteristic values for the slope of the recession curve (α) to determine flow regimes in a karst aquifer.

α^1	Prevailing flow regime
0.0018	Slow-flow or diffuse-flow
0.0058, 0.006	Mixed (intermediate) flow
0.25, 0.13, 0.038	Quick-flow or conduit-dominated flow

[1]Range of characteristic α's from literature (Baedke and Krothe, 2001; Shevenell, 1996; Padilla and others, 1994, Sauter, 1992; and Milanović, 1981a, b).

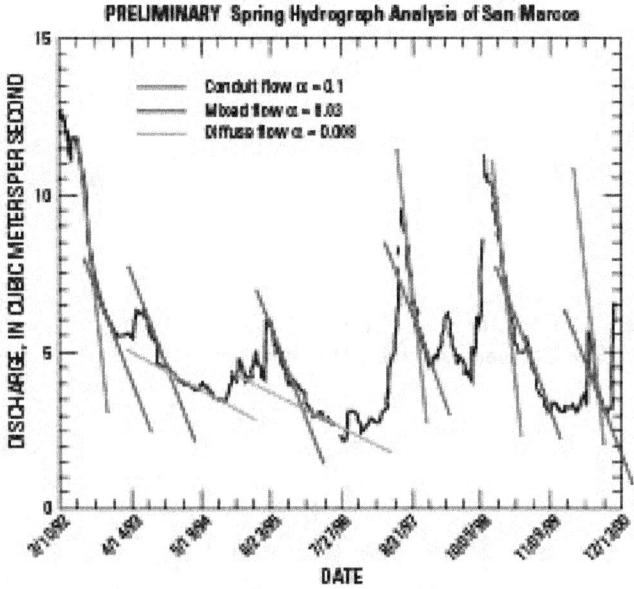

Figure 10. Analysis of spring hydrograph of San Marcos Springs in Texas identifying the conduit, mixed, and diffuse flow regimes of the karst aquifer. (Analysis by Earl Greene, U.S. Geological Survey.)

events, and(or) between waters contributed from the various recharge sources under study, in order to determine mixing proportions (Clark and Fritz, 1997). For example, Lakey and Krothe (1996) used stable isotopes of oxygen ($\delta^{18}O$) and hydrogen (δ^2H) to calculate the mixing proportions of fresh meteoric water and stored ground water discharging from the Orangeville Rise spring basin in south-central Indiana. Studies done by Lee and Krothe (2001) and Trček and Krothe (2002) used sulfate, dissolved inorganic carbon (DIC), $\delta^{13}C$ obtained from DIC, δ^2H, and $\delta^{18}O$ as natural tracers of recharge contributed by matrix ground water, soil water, epikarst water, and fresh meteoric water, and developed three- and four-component mixing models of spring discharge by hydrograph separation (fig. 11). More recently, Doctor and Alexander (2005) used hydrograph recession analysis to identify the flow regimes contributing to spring discharge and then grouped the sampled chemical and isotopic data according to when they were collected with each flow regime. From analysis of these data, water chemistry patterns were identified that were distinctive of each hydrograph-defined flow regime (flood flow, high flow, moderate flow, and base flow).

Precipitation Response Analysis

In well-developed karst aquifers, large springs will act as outlets or drains to the system. The rate of ground-water flow and chemical composition of the spring water is directly related to the basin-scale hydraulic and transport properties of the karst aquifer. Because of the direct connection to surface recharge (sinkholes, sinking streams), karst springs have a wide range of physical and chemical response to precipitation events. Depending on the degree of conduit-to-fracture/matrix coupling, spring hydrographs may show a variable response to recharge events. If there is a high degree of conduit-to-fracture/matrix coupling, the spring will respond in a relatively short time (hours to weeks) to a recharge event, whereas, if this coupling is low, the spring response may take many days or weeks. Knowing how the spring response is related to the recharge events is so important in karst hydrology that much research has been directed toward methods of simulating or predicting this response. Three approaches, linear systems analysis, lumped parameter (statistical modeling), and numerical deterministic modeling, commonly are used to simulate or predict the output function (spring discharge) of a karst system on the basis of the known or measured input function (precipitation pulse).

Linear Systems Analysis

Linear systems analysis has been used in the hydrological sciences for many years to characterize rainfall-runoff relations (Dooge, 1973; Neuman and de Marsily, 1976) and has been used to describe rainfall (recharge)-spring discharge relations in karst systems (Dreiss, 1982; 1983; 1989). The use of a linear method to characterize a nonlinear system (karst ground-water flow) has been justified on a practical basis. First, it is difficult if not impractical to develop a detailed deterministic (numerical) model of ground-water flow in a karst basin because of the difficulty in physically modeling fluid movement in pores, fractures, and conduits. Secondly, the discharge hydrographs of large resurgent springs, like surface-runoff hydrographs, show a response that is directly related to recharge provided by rainfall events. Linear systems modeling will lump many of the complex processes and is useful for describing the karst aquifer.

If a karst system can be conceptualized to act as a linear, time-variant filter, the relation of continuous input (sinkholes, sinking streams, precipitation) can be transformed as continuous output, usually spring discharge (Dreiss, 1982) (fig. 12). The convolution integral below can be used to describe the relation between the output, or spring discharge $y(t)$, and the input, or ground-water recharge $x(\tau)$, and $h(t-\tau)$ is the kernel function (Dreiss, 1982),

$$y(t) = \int_0^t h(t-\tau)x(\tau)d\tau \qquad (8)$$

For two discrete finite series that are causally related, the form of the convolution equation above becomes

$$y_i = \Delta t \sum_{j=0}^{i} x_j h_{i-j} \quad i = 0,1,2,....N \qquad (9)$$

A

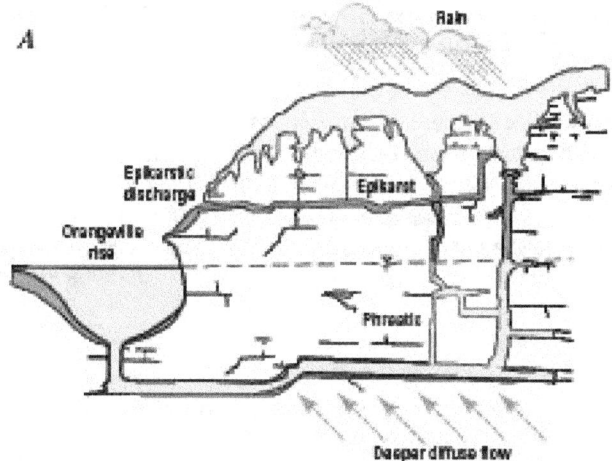

B

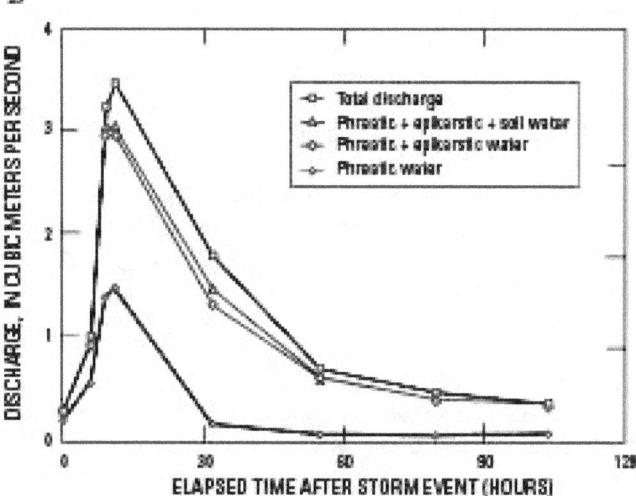

Figure 11.   *A*, Conceptual model of the hydrologic components of the upper Lost River drainage basin in south-central Indiana; *B*, Four-component hydrograph separation curves at Orangeville Rise (from Lee and Krothe, 2001, reprinted from Chemical Geology, copyright 2001, with permission from Elsevier).

for $N+1$ sampling intervals of equal length Δt, where y_i is the mean value of the output during the interval i; x_j is the mean value of the input during the interval j, and h_{i-j} is the kernel function during the interval i-j. Thus, if x_j and h_{i-j} are known, then y_i can be determined directly by convolution. If x_j and y_i can be identified, then h_{i-j} (kernel function) can be determined through deconvolution (Dooge, 1973; Dreiss, 1982, 1989).

Identification of the kernel function from hydrological data is difficult because of the nonlinearities in the hydrological system and errors in the measured data. Then the convolution relation becomes:

$$y_i = \Delta t \sum_{j=0}^{i} x_j h_{i-j} + \varepsilon_i \qquad i = 0,1,2,.....N \qquad (10)$$

where ε_i are the sum of the residual errors. In this case, the identification of the kernel function h_{i-j} is more of an optimization problem and is found by minimizing the sum of the square

errors. Methods of identifying the kernel functions in karst aquifers, in addition to issues involved in defining and working with these functions to predict flow and nonpoint source contamination, are presented in Dreiss (1982, 1989) and Wicks and Hoke (2000).

Wicks and Hoke (2000) applied and expanded the application of linear systems analysis to predict the changes in quantity and quality of water from a large karstic basin. Wicks and Hoke (2000) were able to predict the first arrival time and dispersion of solute discharged from a spring when injected into a specific point (fig. 13).

Long and Derickson (1999) applied a linear systems analysis approach to the karstic Madison aquifer in the Black Hills, South Dakota, to investigate the aquifer's response (head) to an input function. In this instance, stream loss (recharge), which was modeled by using a transfer function, could be related to the total memory length of the karst system (fig. 14). This method could be used as a response-to-recharge event-prediction tool in karst aquifers.

Lumped-Parameter Models

In some karst basins, a linear response (kernel function) cannot adequately simulate the spring outflow. The purpose of lumped-parameter models is to simulate the temporal variations in discharge from springs. When the discharge rate varies continuously and depends on hydrologic input processes of precipitation, sinking streams, evapotranspiration, and infiltration, a model can be developed that produces the output based on some or all of the inputs (Zhang and Bai, 1996). One of the most common nonlinear, lumped-parameter models

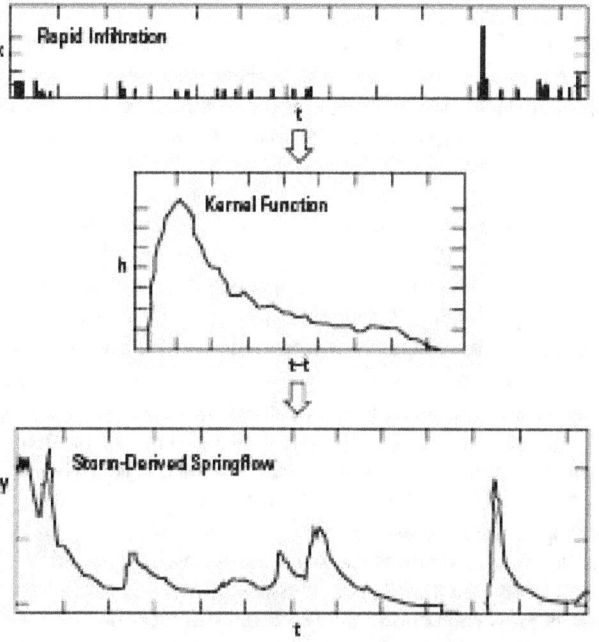

Figure 12.   Linear system analysis of a karst conduit spring showing the recharge-discharge relation (after Dreiss, 1989).

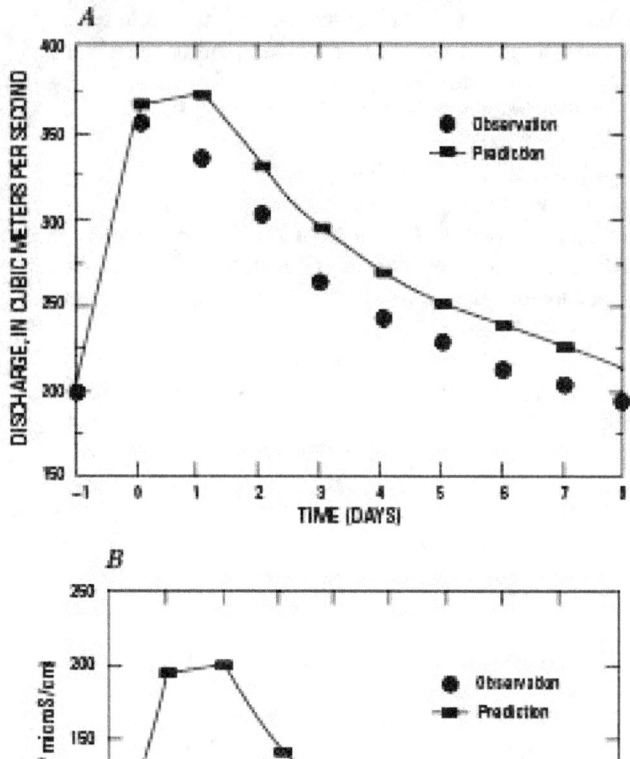

Deterministic (Numerical) Modeling

Numerical modeling has become an important, commonly applied tool for investigating and quantifying many complex hydrogeological relations. However, many technical and conceptual difficulties remain to be solved to facilitate the discretization of conduit geometry or karst basin boundaries, parameterization of rapid- and slow-flow karst components, and simulation of temporal or spatial changes in saturation and flow conditions.

The use of deterministic models is most problematic in quick-flow or conduit-dominated karst systems. Data requirements for parameterization and proper model calibration of conduit-dominated flow are difficult to meet (Teutsch and Sauter, 1991; White, 1999). At present, the technical modeling capabilities and experience base needed to support such applications typically are lacking. Conduit-flow modeling codes are under development that may be of use in studies where the geometry of the conduit system can be fairly accurately mapped (Liedl and others, 2003). Some successes in simulating the effects of conduit flow have been achieved using a modified double-porosity modeling approach (Teutsch and Sauter, 1991) and by embedding high transmissivity zones within the grids of finite-difference or finite-element models (Worthington, 2003; Kuniansky and others, 2001).

Some of the more successful applications of numerical modeling have been in the simulation of spring discharge. Scanlon and others (2003) evaluated two different equivalent porous-media approaches (lumped and distributed parameter) to simulate regional ground-water flow to Barton Springs in the Edwards aquifer, Texas. Both methods worked fairly well to simulate the temporal variability in spring flow

Figure 13.   Predicted and observed (*A*) discharge at Maramec Spring, Missouri; and (*B*) specific conductance times spring discharge water (after Wicks and Hoke, 2000). Reprinted from Ground Water with permission from the National Ground Water Association, copyright 2000.

is the Hammerstein Model, and its use is demonstrated by Zhang and Bai (1996) and Stoica and Soderstrom (1982). The Hammerstein Model is a particularly good, general method for developing a lumped-parameter model for a karst basin. The model uses a least-squares approach to solve for coefficients in the Auto-Regressive Moving Average (ARMA) model and then is used to simulate spring discharge.

Zhang and others (1995) developed a lumped-parameter model to simulate the temporal variations in discharge from Big Spring, Iowa. When precipitation is assumed to be the sole input, the simulated spring discharge matched poorly with the observed spring discharge. The match improved significantly when other variables were added to the model, such as evapotranspiration, infiltration, and snowmelt. Approaches using lumped-parameter models as demonstrated by these authors can be successfully used to simulate spring discharge.

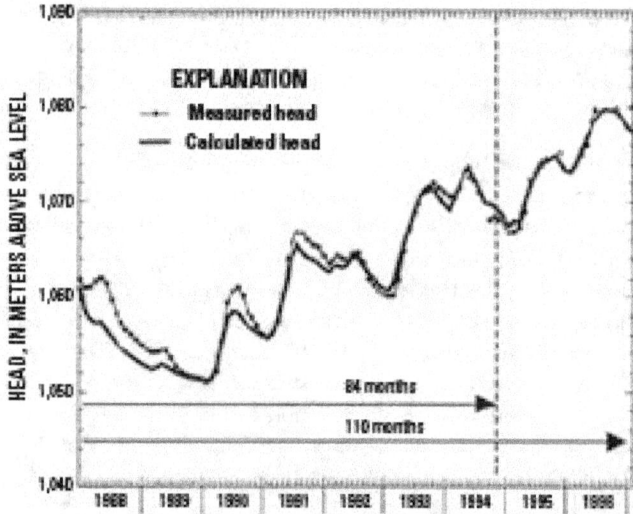

Figure 14.   Calculated and measured head in an observation well, analysis is based on an 84-month time period that was used to predict a 110-month time period (after Long and Derickson, 1999). Reprinted from Journal of Hydrology, copyright 1999, with permission from Elsevier.

(fig. 15). The effect of pumping at a regional scale on spring discharge was best simulated by using a lumped-parameter distribution approach; however, a detailed evaluation of the effect of pumping on water levels and spring discharge required a distributed-parameter approach (Scanlon and others, 2003). Other successful results have been achieved in simulating karst aquifers dominated by slow-flow (diffuse-dominated) components, or in regional-scale studies where the effects of conduit-related heterogeneity can be minimized or neglected.

Deterministic rainfall-runoff models have been used successfully to estimate ground-water recharge and to simulate the hydrologic responses of watersheds in many non-karstic terranes (Beven, 2001; Cherkauer, 2004). Their possible application to karst hydrology studies seems promising but has received little attention thus far. As with deterministic ground-water models, a variety of technical and conceptual difficulties currently limit the use of these models. Larger, regional-scale modeling may be less problematic (Arikan, 1988). Available rainfall-runoff modeling codes such as TOPMODEL (Beven and Kirkby, 1979) and PRMS (Leavesley and others, 1983) are not well suited to dealing with issues related to internal drainage by sinkholes or routing of subsurface flow through conduits. Various "workarounds" such as filling and smoothing sinkhole depressions, or artificially inflating the volume or storage capacities of sinkholes, have been used experimentally to achieve model calibration (Campbell and others, 2003). These approaches have not been very successful, however, and have resulted mostly in models that do not accurately represent the physical hydrogeologic conditions in the karst basin or simulate the full range of observed flow conditions and hydrologic responses. Additional research aimed at improving the conceptualization and parameterization of karst flow systems in rainfall-runoff models is needed and would be beneficial.

Water Tracing with Fluorescent Dyes

Water tracing with fluorescent dyes is a particularly useful tool for investigating water fluxes in karst flow systems because dye-tracer tests can be used to obtain direct information about flow direction, velocity, and other hydraulic characteristics in conduits between specific points of focused recharge and discharge. Fluorescent dyes are organic chemicals that absorb light from the ultraviolet part of the spectrum, are molecularly energized, and emit light at a longer wavelength range (Käss, 1998). As described by Smart and Laidlaw (1977) and Field and others (1995), an ideal water tracer is one that (1) is easy to introduce into the aquifer or flow system; (2) travels at or near the flow rate of water; (3) is relatively conservative—that is, not easily lost through sorption; (4) is stable with regard to water chemistry; (5) is easily detectable at low concentrations; and (6) has little or no toxicity to humans or aquatic organisms and poses no long-term intrinsic threat to the environment. As a group, fluorescent

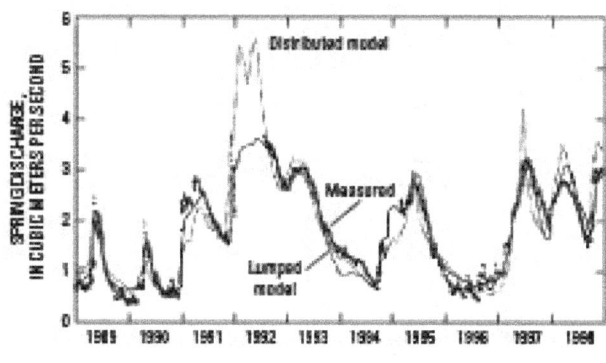
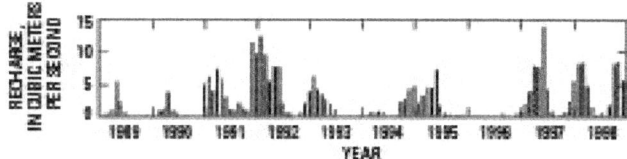

Figure 15. Simulation of Barton Spring discharge in the Edwards aquifer, Texas, using a lumped and distributed parameter approach (after Scanlon and others, 2003). Reprinted from Journal of Hydrology, copyright 2003, with permission from Elsevier.

dyes possess almost all of these characteristics and, as such, they are widely used and popular choices for artificial tracer tests in karst studies.

Several of the fluorescent dyes most commonly used in water-tracing investigations in karst are listed in table 6. The fluorescent characteristics, detection limits in water, and sorption tendencies of these dyes are provided in table 7. Most of the individual dyes listed are members of a family or group of dyes that vary slightly in chemical structure and have overall similar fluorescent properties. The xanthene dyes are a large group that exhibit fluorescence in the green to orange wavelengths of the visible light spectrum (Käss, 1998) and includes such well-known tracer dyes as sodium fluorescein (also known as uranine), which fluoresces in the green wavelength band (500–570 nanometers [nm]), and Rhodamine WT, which fluoresces in the yellow-orange wavelength band (570–590 nm). Another large group of tracers is the stilbenes, or optical brighteners—compounds that technically are not dyes but are whitening agents—which fluoresce in the violet-blue wavelength range of the visible light spectrum (380–500 nm). Trade names of individual dyes may vary by manufacturer or supplier, so it is advisable to refer to a specific tracer dye using the Color Index (CI) generic name and the Chemical Abstracts Service (CAS) identification number (Field and others, 1995).

Successful use of fluorescent dyes in water-tracing studies requires at least a general working knowledge of the physical and chemical properties of individual dyes, and the conditions and limitations involved in their use. For example, fluorescence is pH and temperature sensitive; however, different dyes have different ranges of sensitivity and response to these properties. Certain dyes, such as sodium fluorescein, are particularly photosensitive, whereas others, such as Rhodamine WT, are not.

Table 6.   Some commonly used fluorescent dye types, their dye names, and their respective Color Index and Chemical Abstracts Service (CAS) number (from Field, 2002b).

Dye type and common name	Color index generic name	CAS No.
Xanthenes		
sodium fluorescein	Acid Yellow 73	518-47-8
eosin	Acid Red 87	17372-87-1
Rhodamines		
Rhodamine B	Basic Violet 10	81-88-9
Rhodamine WT	Acid Red 388	37299-86-8
Sulpho Rhodamine G	Acid Red 50	5873-16-5
Sulpho Rhodamine B	Acid Red 52	3520-42-1
Stilbenes		
Tinopal CBS-X	Fluorescent Brightener 351	54351-85-8
Tinopal 5BM GX	Fluorescent Brightener 22	12224-01-0
Phorwite BBH pure	Fluorescent Brightener 28	4404-43-7
Diphenyl Brilliant Flavine 7GFF	Direct Yellow 96	61725-08-4
Functionalized polycyclic aromatic hydrocarbons		
Lissamine Flavine FF	Acid Yellow 7	2391-30-2
pyranine	Solvent Green 7	6358-69-6
amino G acid	--	86-65-7

Table 7.   Emission spectra and detection limits for dyes in water (modified from Field, 2002b).

[nm, nanometer; %, percent; µg/L, micrograms per liter]

Dye name	Maximum excitation λ (nm)	Maximum emission[1] λ (nm)	Fluorescence intensity (%)	Detection limit[2] (µg/L)	Sorption tendency
sodium fluorescein	492	513	100	0.002	Very low
eosin	515	535	18	0.01	Low
Rhodamine B	555	582	60	0.006	Strong
Rhodamine WT	558	583	25	0.006	Moderate
Sulpho Rhodamine G	535	555	14	0.005	Moderate
Sulpho Rhodamine B	560	584	30	0.007	Moderate
Tinopal CBS-X	355	435	60	0.01	Moderate
Phorwite BBH Pure	349	439	2	--	--
Diphenyl Brilliant Flavine 7GFF	415	489	--	--	--
Lissamine Flavine FF	422	512	1.6	--	--
pyranine	460[3]	512	18	--	--
	407[4]	512	6	--	--
amino G acid	359	459	1.0	--	--
sodium napthionate	325	420	18	0.07	Low

[1] Approximate values only.

[2] Typical values for tracer detection in clean water using spectrofluorometric instrumentation.

[3] For pH greater than or equal to 10.

[4] For pH less than or equal to 4.5.

In addition, fluorescent dyes have varying ranges of reactivity with geological materials such as clays and other silicate minerals (Käss, 1998; Kasnavia and others, 1999). These and other important physiochemical characteristics always need to be considered prior to use. Useful references include Smart and Laidlaw (1977), Mull and others (1988), and Käss (1998). Although toxicity generally is not a great concern with most of the fluorescent dyes commonly used for water-tracing studies, this and other possible environmental concerns are reviewed by Smart and Laidlaw (1977) and Field and others (1995).

Dye-Tracer Test Objectives and Design

Dye-tracer testing is a versatile method that can be employed in a number of ways by using various combinations of field and laboratory techniques that can be tailored to fit the specific objectives, context, and scale of the investigation (Smart, 2005). The basic goal of any dye-tracer test is to create a detectable fluorescent signal in water that can be positively identified as originating from the injected tracer dye and that can be interpreted in a manner needed to achieve the planned objectives of the test. Careful planning and execution of dye-tracer tests is essential so that positive, understandable results are obtained from each test—that is, from each injection of a tracer dye (Quinlan, 1989). If a dye is injected and not detected, the investigator may be faced with difficult and often costly decisions that may include whether or not to repeat the test, to change the type or amount of dye injected, to use additional monitoring sites, to conduct monitoring for a longer period of time, or to evaluate the sensitivity of the analytical method being used to detect the presence and (or) concentration of the injected dye.

In practice, dye-tracer tests generally are categorized as either quantitative or qualitative, depending largely on type of monitoring used and the data to be collected and interpreted (Jones, 1984a,b; Mull and others, 1988; Smart, 2005). Fully quantitative dye-tracer tests require accurate measurement of the amount (mass) of tracer dye injected, the discharge from the spring or aquifer during the test, and the concentration or total mass of tracer dye resurging from the aquifer. Quantitative dye-tracer tests primarily are used to obtain information about the time-of-travel and breakthrough characteristics of the tracer dye—which are important to contaminant-related studies—and to investigate karst conduit structure and flow properties (Field and Nash, 1997). Provided that discharge is measured simultaneously with tracer concentration at all dye-resurgence points, tracer mass recovery can be determined and used to make reliable estimates of conduit hydraulic properties including mean residence time, mean flow velocities, longitudinal dispersion, and storage (Field, 2002b).

In contrast to quantitative dye-tracer tests, qualitative dye-tracer tests are those that require only a determination of positive or negative resurgence of injected tracer dye at monitoring sites used for the test (Jones, 1984a). Qualitative dye-tracer tests are usually conducted to identify flow connections between focused points of recharge and discharge (for example, a sinkhole and a spring), thereby helping to delineate the trajectories of subsurface flow paths and to estimate an approximate maximum time-of-travel (based on the sampling interval used). Discharge data typically are not collected, and the actual concentrations of dye resurging in water at each monitoring site may or may not be determined, depending on the analytical methods used. Monitoring for these types of tracer tests often is accomplished by using passive detectors made of an adsorptive media such as granular activated charcoal to trap the tracer dye.

Planning required for a dye-tracer test typically involves a careful review of available hydrogeologic information, selection of dye-injection and dye-monitoring sites, an assessment of ambient fluorescence and hydrologic (flow) conditions, and selection of a method or methods to be used for dye monitoring and detection that is appropriate for the objectives and category of tracer test (that is, quantitative or qualitative). Qualitative dye-tracing and quantitative dye-tracing methods are not mutually exclusive, and the two methods often are used in combination in many karst studies (Quinlan, 1989).

In common practice, quantitative dye-tracer tests often are conducted after subsurface flow routes have been identified between specific input points and discharge points by qualitative dye-tracer tests.

During any dye-tracer test, it is important that all potential dye-resurgence sites be identified and monitored to ensure that complete recovery of tracer dye is achieved. The results of previously conducted dye-tracer tests are very useful in the planning of subsequent tests, so every effort needs to be made during the planning phase to identify and review existing dye-tracer test information. For studies intended to delineate subsurface flow paths or karst basin boundaries, previous dye-tracer test results, estimates of unit-base flow of local springs, and other types of available hydrogeologic mapping data, are helpful in establishing the size and boundary of the study area required for monitoring. If few subsurface flow routes have been dye-traced and karst basin boundaries have been only partly delineated, or are not known, it may be necessary to monitor many springs in the study area, even those thought to be improbable resurgence sites, to ensure detection of the injected tracer dye.

Information about local ground-water flow directions and hydraulic gradients is extremely useful in the planning and the interpretation of dye-tracer tests. Therefore, if suitable water-level or potentiometric-surface maps are not available, it is wise to conduct an inventory and synoptic water-level survey of wells in the study area prior to initiation of a dye-tracer test. For many studies, selected wells need to be incorporated in addition to springs and streams as potential dye-monitoring sites. A field reconnaissance also needs to be done prior to implementation of any dye-tracer test. This is often a necessary and underappreciated aspect of the planning process. During the reconnaissance, potential dye-injection and dye-monitoring sites can be located and inspected to identify any logistical issues that may affect the implementation of the planned tracer test. Springs identified on published topographic maps often are inaccurately located, and the number of spring outlets plotted on a topographic map of a given area can be underrepresented as well (Quinlan, 1989). A thorough spring inventory needs to be conducted as part of the tracer-test planning process by searching existing databases; by walking, wading, or boating along surface stream reaches within the selected study area; by consulting aerial photographs; and by interviewing local landowners.

Because of the rapid temporal changes in hydraulic gradients, flow velocities, and flow directions typical of many conduit-dominated karst aquifers, the results obtained during a specific dye-tracer test are, strictly speaking, representative only of the flow conditions existing at the time of the test. For this reason, some consideration needs to be given during the planning phase as to whether additional dye-tracer tests need to be conducted during specific high- or low-flow conditions. For practical reasons, most dye-tracer tests are conducted during moderate- or base-flow conditions. During low-flow conditions, greater losses of tracer dye, and longer resurgence times, can be expected than at high-flow conditions, because of sorption, low-flow velocities, and storage of dye in hydraulic "dead-zones." Different issues may occur during high- or flood-flow conditions. Injected tracer dyes may become too diluted and resurge in springs at concentrations below detection limits, the increased turbidity may interfere with dye monitoring and detection, physical access to dye-monitoring sites may be hindered, and in-situ dye-monitoring equipment may be damaged by flooding. In addition, hydraulic damming of conduits caused by flooded streams may temporarily halt or delay the resurgence of tracer dyes.

Dye Injection

Dyes are typically injected as a "slug" of known weight, volume, or mass (fig. 16). A principal cause of negative or inconclusive dye-tracer test results is the injection of an insufficient quantity of dye into the aquifer (Quinlan, 1989; Field, 2003). Proper determination of the amount of dye to inject also is needed to ensure that dye resurges at detectable but not unacceptably high concentrations, particularly in public or private water supplies, and that residual storage of dye in the aquifer is minimized. Because of concerns about the possible formation of the carcinogen diethylnitrosamine resulting from the use of Rhodamine WT dye (Steinheimer and Johnson, 1986), the USGS adopted a policy that the concentrations of Rhodamine WT should not exceed 10 µg/L during tracing tests of surface waters near public water intakes (Water Resources Division Memorandum No. 66.90 and 85.82). In a review of toxicity and other environmental data, Field and others (1995) suggested that the resurgent concentration of most commonly used tracer dyes should not exceed 1 to 2 mg/L (1,000–2,000 µg/L) for more than 24 hours at a point of ground-water withdrawal or discharge. These concentration limits, while desirable, may not always be possible to achieve because of the unpredictability of subsurface flow routes and field variables that affect the rate of transport, dispersion, and subsequent concentration of dye discharged through conduits.

Historically, a variety of equations have been devised to estimate the quantity of dye needed for tracer test injections, based largely on distance to the anticipated resurgence point and or estimated ground-water flow velocities. Most of these are difficult to apply in practice and do not provide a means for the investigator to predict and manage the resurgent concentration of tracer dye. These shortcomings are addressed in methods devised by Field (2003) and by Worthington and Smart (2003).

The Efficient Hydrologic Tracer-Test Design (EHTD) method by Field (2003) includes a computer program that estimates the amount of dye needed for injection and provides forward modeling capability needed to predict tracer-breakthrough curve characteristics and the time intervals needed for effective sampling of the passage of the dye pulse—information that is important to planning quantitative tracer tests. The EHTD method calculates the amount of dye needed for injection by using various forms of the advection-dispersion equation for open-channel flow, closed-conduit flow, and flow through porous equivalent media. The program enables the user to designate the mass of tracer dye to be injected and an injection flow rate. For open-channel and closed-conduit flow conditions, the EHTD method requires the following input values: (1) discharge at the sampling station (spring), (2) estimated longitudinal distance from the dye-injection site to the anticipated resurgence site, (3) estimated cross-sectional area of the discharge point (that is, spring or stream cross-sectional area), and (4) a sinuosity factor applied to straight-line estimates of the distance between a dye injection and potential dye resurgence site.

The method proposed by Worthington and Smart (2003) relies upon the empirically derived equations:

$$M = 19 \ (LQC)^{0.95} \qquad (11)$$

and

$$M = 0.73 \ (TQC)^{0.97} \qquad (12)$$

where

M is mass of tracer dye injected (grams/meter³),

L is anticipated distance between the injection site and the anticipated primary resurgence site (meters),

Q is discharge at the anticipated resurgence (meters³/second),

C is peak tracer concentration at the anticipated resurgence (grams/meter³),

and

T is traveltime as determined from prior tracing-test results (seconds).

Using either equation, the investigator can select a target concentration desired for resurging tracer dye and solve to determine the required amount (mass) of dye needed for injection.

In practice, dye injection is best accomplished at locations that provide rapid, direct transport of the tracer into conduits, thus minimizing loss of dye through photochemical decay, sorption, or other field conditions. Open-throated swallets in sinkholes and the swallow holes of sinking streams are ideal sites. In the absence of naturally occurring runoff (inflow), dyes can be injected into a stream of potable water discharged from a tanker truck or large carboy. In general, 300 to 500 gallons of water are a minimum quantity needed for dye injection, and quantities of 1,000 gallons or more are preferable. Approximately one-half of the water is used to initiate flow into the swallet prior to dye injection. This is done to test the swallet's drainage capacity, to initiate flow, and to flush the flow path to minimize losses to sorption. The remainder of the water is discharged after the dye is injected as a "chaser." Under most conditions, this technique does not substantially change the naturally occurring flow conditions or alter hydraulic heads in the aquifer.

A

C

B

Figure 16.   Dye injections: *A*, sodium fluorescein injection into collapse sinkhole formed in a pond (stream of water is outflow from a settling pond at a public supply water-treatment plant); *B*, Rhodamine WT injected into sinking stream; *C*, injection of Rhodamine WT into a water-level observation well (photographs by Charles J. Taylor, U.S. Geological Survey).

Several practical tips regarding dye injection that may be useful to consider during the planning of dye-tracer tests include the following:

- Open-borehole wells or screened wells can be used as dye injection sites. Although a pre-flush is not necessary prior to injecting the dye, it is advisable to conduct a falling-head slug test in order to test the hydraulic connection with the aquifer and to estimate the local hydraulic conductivity and rate of discharge of tracer from the well. After injecting dye, it is necessary to flush the dye from the well using several borehole volumes of potable water. It is important to control the volume and rate of water inflow during the flush to prevent the well from overflowing with dye-laden water.

- In locations where available sinkholes do not contain open swallets, dye injection may be accomplished by drilling a temporary injection well. Ideally, these injection wells would be drilled to intercept fractures or solutional openings in the bedrock; however, successful

injection into the epikarst may be accomplished by completing the well at the top of the karstic bedrock. Aley (1997) discusses in detail the issues involved in conducting dye-tracer tests through the epikarst zone.

- Dye injections also can be made through the epikarst by excavating a pit into the soil; however, dye losses may be significant, and the quantity of dye used for the injection usually must be increased several times above the "normal" dosage amount.

- The injection of dye into flooded sinkhole depressions or swallets choked with sediment generally is not advisable, particularly for quantitative tracer tests. Unless there is evidence that drainage through the regolith into the subsurface is relatively rapid, excessive loss of the tracer dye may be incurred. If necessary, swallets that are partly choked with sediment may be cleared out with a shovel or backhoe and pre-tested for drainage capacity prior to an attempted injection of dye.

- A slug injection may not be an effective means of attempting to trace flow from a losing stream—that is, into a stream not fully diverted underground by swallow holes—because too much of the dye may be flushed downstream before it infiltrates the subsurface. Under such conditions, dye injection may be more effectively accomplished by using the continuous-injection technique described by Kilpatrick and Cobb (1984). A recent paper by Field (2006) specifically examines the problem of conducting dye-tracer tests from losing streams.

Dye Monitoring and Detection

As dictated by the tracer-test objectives and the resources available, a variety of methods can be used to monitor for and detect the resurgence of injected tracer dyes, including direct visual observation, fluorometric analysis of discrete water samples or eluant obtained from granular activated charcoal detectors, and in-situ continuous-flow fluorometry. Dyes often can be visually detected in water at parts-per-million concentrations, whereas some method of fluorometric analysis is needed to detect dyes at subvisual concentrations. Three types of fluorometers are commercially available: scanning spectro-fluorophotometers, filter fluorometers, and in-situ submersible fluorometers. Each of these instruments operates essentially by selectively measuring the fluorescent intensity of a sample (for example, water) that has been selectively excited (Duley, 1986). The selective range of light wavelengths used to excite the sample is called the excitation spectrum, and the selective range of light wavelengths that are measured by the fluorescent intensity is called the emission spectrum (Käss, 1998). Depending on which type of the three instruments is used, common tracer dyes can be detected in water at concentrations as low as parts per trillion—although environmental factors usually limit unequivocal detection to the range of parts per billion or greater (Smart and others, 1998).

Scanning spectrofluorophotometers are research-grade laboratory instruments that use a system of monochromators, diffraction gratings, and bandwidth slits to scan across user-selected excitation and(or) emission spectra at selected bandwidth intervals. These instruments are exceptionally sensitive—dyes often can be detected in the parts-per-trillion range—and enable precise characterization of the various sources of fluorescence in a sample. One advantage of these instruments is that they can be used to do synchronous scanning, a technique in which the excitation and emission monochromators are scanned together at a fixed wavelength difference determined by the separation (in nanometers) between the excitation and emission peaks for the dye(s) of interest (Duley, 1986; Rendell, 1987). For most xanthene dyes, this distance is approximately 20 to 25 nanometers (Käss, 1998). The synchronous scanning technique is useful for analyzing unknown mixtures of fluorescent solutes having various excitation wavelengths because it provides a spectral "fingerprint" for each solute present and therefore can be used to

identify the presence of multiple tracer dyes in a single sample (Käss, 1998) (fig. 17). The tradeoff in using these instruments is that they require a good working knowledge of relatively specialized fluorescence spectroscopy methods and the use of rigorous quality-control methods, which can be quite time and labor intensive.

Filter fluorometers, such as the Turner Designs Model 11, are versatile instruments that can be set up to work under field or laboratory conditions, to analyze discrete samples of water or eluant or be used with flow-through cells and pumps to continuously monitor the change in fluorescence due to the passage of a dye pulse in water. These instruments have user-exchangeable glass filter kits that transmit light in the excitation and emission wavelengths of fluorescein or rhodamine dyes and are capable of identifying dyes at concentrations in the parts-per-billion range. Because the excitation and emission wavelengths are fixed by the set of filters installed, only one tracer dye at a time can be identified in a sample. Filter fluorometers are dependable "workhorse" instruments widely used for dye-tracer studies in karst. As discussed by Smart and others (1998), however, there are a number of practical constraints on the use of filter fluorometers, the most serious of which is that the sensitivity of these instruments may be adversely affected by ambient fluorescence so that it is possible to obtain apparently significant fluorescent readings—indicating positive detection of dye—with a particular filter set when the dye of interest is not actually present. Such false positive readings result from the presence of other fluorescent solutes having fluorescence spectral properties that overlap those of the tracer dye of interest (Smart and others, 1998).

Recently, manufacturers such as Turner Designs and Yellow Springs Instruments have introduced submersible fluorometers that can be used for in-situ continuous-flow monitoring of either sodium fluorescein or Rhodamine WT dyes. The filters needed for detection of each dye are preinstalled by the manufacturer in an optical probe assembly. These instruments include internal data loggers capable of recording thousands of data values in nonvolatile flash memory and simultaneously collect temperature and turbidity data as needed to correct the recorded fluorescent intensity values. Continuous-flow fluorometry conducted with either filter fluorometers or submersible fluorometers provides significant advantages for quantitative dye-tracer tests because highly resolved dye-breakthrough curves can be obtained whose properties are not as affected by sampling biases or by insufficient sampling frequency (Smart, 1998). Quality control also may be considerably improved because the water is analyzed directly without excessive sampling and handling activities that potentially increase the chances of sample contamination or degradation. These advantages sufficiently outweigh any potential loss of spectral precision (Smart and others, 1998).

For quantitative analysis, all fluorometers must be calibrated so that the concentration of dye is determined by the fluorescent intensity of the sample measured relative to that of dye-concentration standards. Standards are prepared from the tracer-dye stock solution by using gravimetric and serial dilution techniques in the manner described by Wilson and others (1986) or Mull and others (1988). Turner Designs offers a number of application notes that can be downloaded or ordered from their Internet website (*http://www.turnerdesigns.com*) that describe in detail the procedures involved in the preparation of dye standards and the calibration of filter fluorometers. The preparation of dye-concentration standards is a critical procedure and needs to be undertaken with great care. Errors introduced into the standards preparation process will adversely affect instrument calibration and, for quantitative tracer tests, may result in serious mass-balance errors (Field, 1999b). Adjustments to account for the percent actual tracer in powdered and liquid

dyes, and also for specific gravity in liquid dyes (Field, 1999b, 2002b) (table 8) must be factored in during the calculation of dye standard concentrations. Dye-concentration calibration curves typically are made by using a logarithmic distribution of dye-concentration standards ranging over two or three orders of magnitude (Alexander, 2002).

In practice, error in dye detection and(or) determining dye concentration in water is dominated by issues of ambient (background) fluorescence, loss of tracer (that is, due to adsorption or photodegradation), and improper sampling frequency (Smart, 2005). Ambient (background) fluorescence is probably the largest single source of systematic error in dye

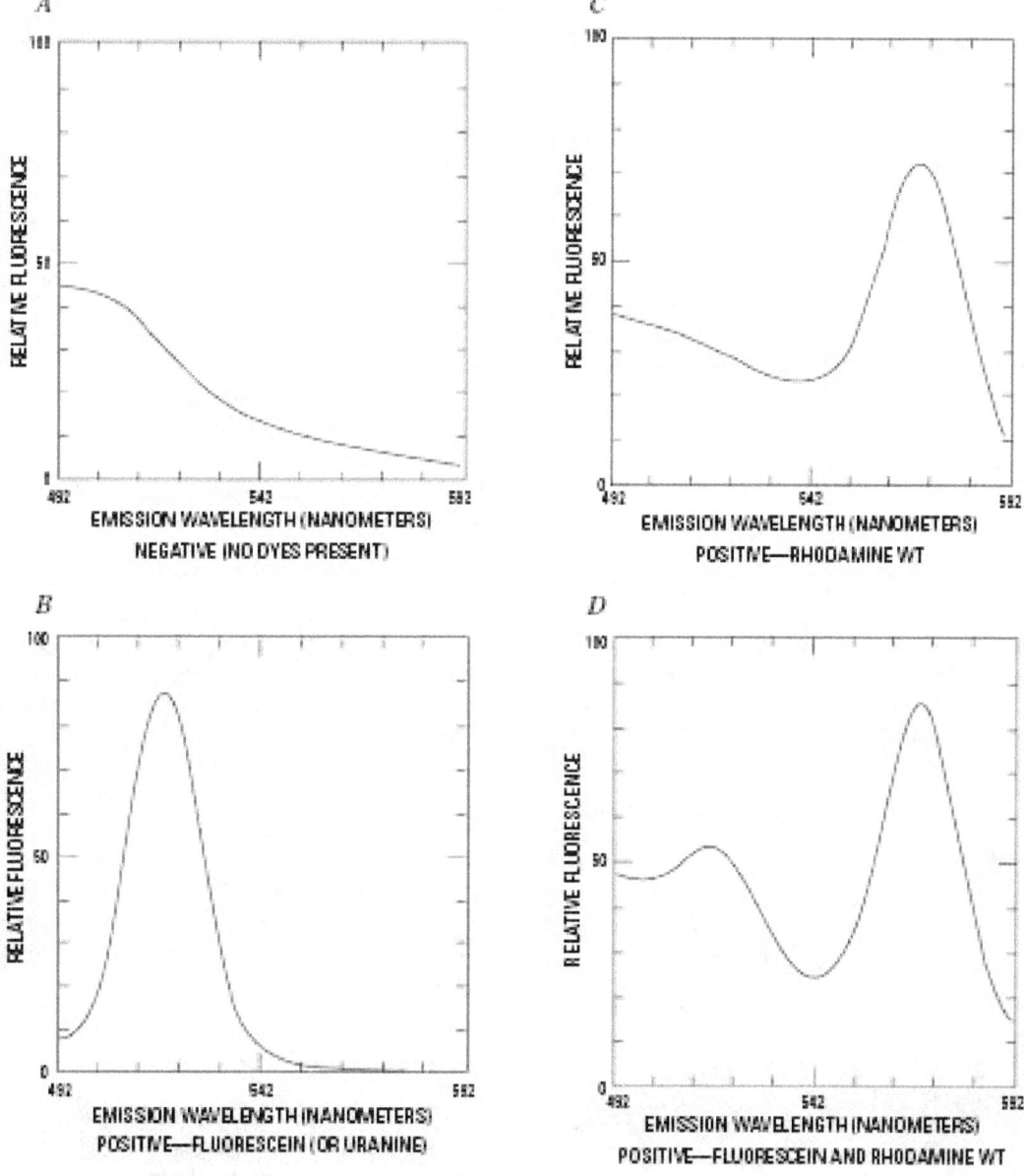

Figure 17.   Dye spectral "fingerprints" obtained from use of the synchronous scanning method: *A*, no fluorescent tracer dye is present; *B*, sodium fluorescein (or uranine) tracer dye is present; *C*, Rhodamine WT tracer dye is present; *D*, sodium fluoresein and Rhodamine WT tracer dyes are present (after Vandike, 1992).

tracing and must be carefully assessed prior to initiation of any (qualitative or quantitative) dye-tracer test. For quantitative tests, ambient background levels occurring in the range of the emission peak of the tracer dye being used must be subtracted to accurately calculate dye concentrations. Often, the choice of dye selected for a particular tracer test is influenced by the presence and level of fluorescent intensity of ambient fluorescent interferences. Potential sources of interference with tracer dyes include naturally occurring humic and fulvic acids, certain species of algae, petroleum hydrocarbons, optical brighteners discharged in septic or treated waste-water effluent, automotive antifreeze chemicals (a widespread source of fluorescein), and hundreds of other dyes and organic chemicals used in industrial, commercial, and household products (Käss, 1998). In general, ambient background interferences typically are more problematic for optical brighteners and for xanthene dyes that fluoresce in the blue-green spectral wavelengths (Käss, 1998), and less problematic for xanthene dyes that fluoresce in the yellow-orange spectral wavelengths (Smart and Karunaratne, 2002).

One important point to consider is that the timing, duration, and intensity of fluorescence can vary considerably, depending on its sources, during the period over which ambient fluorescence is being monitored (Smart and Karunaratne, 2002). For this reason, it is advisable to conduct background monitoring for a period of at least several days or weeks immediately prior to initiating any dye-tracer test. It is also advisable to contact local, state, and Federal water-resources agencies at this time to determine whether or not other tracer tests are in progress or have recently been completed in order to be aware of, and avoid, interference and potential cross contamination with a previously injected tracer dye.

Table 8. Percent active tracer, and specific gravity, measured for some commonly used fluorescent dye tracers. These may vary from batch to batch and should be determined for the specific lot of dye being used for mass-balance calculations attempted during quantitative dye-tracing tests (Field, 2002b).

Color index generic name	Powder dye (%)	Liquid dye (%)	Specific gravity (g cm⁻³)
Acid Blue 9	74.0*	37.0	--
Acid Red 52	90–90.2	18.0	1.175
Acid Red 87	86.0	26.0	--
Acid Red 388	85.0**	17.0	1.160
Acid Yellow 73	60.0	30.0	1.190
Basic Violet 10	90.0	45.0†	--
Fluorescent Brightener 351	60.0	--	--

Values listed are equal to within 5.0 percent.

*Acid Blue also is sold with a Food, Drug and Cosmetic (FD&C) purity equal to 92.0%.

**Acid Red 388 is not commercially available in powder form.

†Basic Violet 10 as a liquid is mixed with glacial acetic acid.

Note: The values listed are specific to one manufacturer; crude dye stocks can and will vary significantly with manufacturer.

Because of ambient fluorescence and other analytical variables involved in fluorometry, there may be some subjectivity and difficulty in assessing the results of a single sample analysis—that is, the question might be asked "Is there enough of a change in fluorescent intensity to indicate that the tracer dye has been detected?" These decisions are made somewhat more objectively if minimum threshold concentrations or fluorescent intensity values are established by statistical methods or some other means to ensure that the fluorescence intensity or concentration measured in a sample is sufficiently higher than background to provide a high confidence level that dye was positively detected. From the literature, it appears that many researchers apply an arbitrary 10:1 signal-to-noise ratio for fluorescent intensity or dye concentration measured at the expected emission peak of the tracer dye as a minimal threshold for reporting the positive detection of dye in a sample (Smart and Simpson, 2001). A number of analytical and data post-processing techniques also have been devised in an attempt to enhance the detection of tracer dyes, particularly when working under "noisy" fluorescent background conditions (Smart and others, 1998; Smart and Smart, 1991; Lane and Smart, 1999; Tucker and Crawford, 1999). More recently, the use of advanced spectral analysis techniques has been explored as a means of better distinguishing tracer dyes from ambient background fluorescence (Alexander, 2005).

In general, caution needs to be used when making a determination that breakthrough and detection of an injected tracer dye has occurred based on only one "positive" sampling result. Evidence of dye breakthrough and detection is more conclusive if repeated positive detections are obtained, particularly where these results demonstrate a change in dye concentrations or fluorescent intensity that is indicative of the passage of a dye pulse and subsequent return to ambient fluorescent conditions. This is a principal reason that some researchers strongly recommend the use of quantitative dye-tracer tests methods, which include high-frequency sampling using automatic water-samplers or continuous-flow fluorometry, whenever possible (Field, 2002b; Kincaid and others, 2005). The passage of a dye pulse, however, also can be conclusively demonstrated by changes in fluorescent intensity or equivalent dye concentration obtained during qualitative dye-tracer tests using passive samplers, provided that a sufficiently high sampling frequency is used. Where questionable or inconclusive dye-tracer test results are obtained, it is advisable to review the tracer test design—particularly the methods used for monitoring and detection—and repeat the test using a different tracer dye.

Use of Charcoal Detectors

As previously indicated, the use of passive detectors containing granular activated charcoal is a popular method of monitoring for dye resurgence during qualitative tracer tests. The detectors typically are constructed of fiberglass screen, nylon netting, or a similar material, fashioned into packets that contain several grams of charcoal sampling media. The size and shape of the packets and the amount of charcoal used in them is not particularly critical—the only requirements are that detectors be relatively durable, securely retain the activated granular charcoal, and allow water to flow easily and evenly through the packets. The detectors, or "bugs", generally are used in rapidly flowing water in streams and springs suspended above the substrate using a wire-and-concrete or wire-and-brick anchor. The detectors also can be staked or pinned directly into the streambed in very shallow water, and they can be easily suspended in monitoring or water-supply wells using, for example, monofilament line, snap-swivels, and steel shot-weights sold as fishing tackle. A common practice is to use detectors at all anticipated resurgences and exchange the detectors at 2 to 10 day intervals throughout the duration of the test (Quinlan, 1989). As a practical matter, it is generally inadvisable to leave detectors in the field longer than 10 days because of physical degradation that can diminish the adsorptive capability of the charcoal.

The principal advantage of using charcoal detectors is their economy and relative ease of use for ground-water reconnaissance studies, for simultaneous monitoring of many potential dye resurgence sites, and for mapping of conduit-flow paths or karst basin boundaries in areas where these are primarily or completely unknown. The detectors are relatively easy to conceal, thus minimizing the potential for disturbance and vandalism; and they are inexpensive, most of the cost being associated with their use onsite, collection, and analysis (Smart and Simpson, 2001). Handling, storage, and transportation requirements used in the exchange of detectors are not particularly critical with the exception of simple procedures needed to eliminate the potential for misidentification of detectors or cross contamination during handling and to prevent degradation of the dyes adsorbed by the charcoal (Jones, 1984a).

Another benefit of using charcoal detectors is their ability to concentrate dye at levels 100 to 400 times greater than the concentration of dye resurging in water, thereby helping to increase the probability of a positive detection of dye at distant monitoring sites (Smart and Simpson, 2001, 2002). To expel adsorbed dye, a few grams of charcoal are removed from a detector packet and eluted in an alkaline-alcohol solution. Two popular eluants include the so-called "Smart solution" (Smart, 1972), prepared by mixing 1-propanol, distilled water, and 28 to 30 percent ammonium hydroxide in a 5:3:2 ratio, and a solution of 70 percent 2-propanol and 30 percent deionized water saturated with sodium hydroxide (Alexander, 2002). Upon mixing, the solution separates into a lighter (saturated) and denser (supersaturated) liquid and it is the lighter phase that is decanted off and used as the actual eluant. Other eluant formulas may be chosen to enhance the elution of specific tracer dyes (Kass, 1998). Any prepared eluant always needs to be scanned as a blank before actual use (Alexander, 2002). Generally, 1 hour of elution is needed before fluorometric analysis can be done, although different dyes have different optimal elution times in various eluants (Smart and Simpson, 2001). Elution may be done with wet or dry charcoal; for

longer term storage, however, charcoal samples need to be completely dried to prevent microbial degradation of the adsorbed dye.

If dye elutes from the charcoal below visible concentrations, an aliquot of the eluant can be removed for analysis by using either a filter fluorometer or a scanning spectro-fluorophotometer (Smart and Simpson, 2002). As with water samples, the relative concentration of tracer dye in the eluant is determined by the fluorescent intensity, or the area of the spectral peak, measured at the emission wavelength of the dye (Jones, 1984b; Smart and Simpson, 2002). Because fluorescence for most tracer dyes is pH-dependent, the emission wavelengths for dyes in alkaline-alcohol eluants generally are shifted several nanometers relative to the emission wavelengths reported for dyes in water samples at or near neutral pH (Kass, 1998), and the emission peak characteristics and calibration curves obtained by fluorometric analysis may vary for different eluant formulations.

Although the use of charcoal detectors is relatively easy and has many potential benefits, the method is not without its shortcomings. Variables in the field, differences in the adsorptive efficiency of charcoal with various tracer dyes, complexities associated with the adsorption-desorption process of organic solutes on charcoal, and other variables introduced as a result of processing in the laboratory, preclude any determination of the actual concentrations of tracer that resurged in water and the replication of analytical results obtained from eluted charcoal (Smart and Simpson, 2001, 2002). The amount of dye concentrated on the detectors is a factor of the rate of flow through the detectors, the total surface area exposed to the dye, and of the length of time of the exposure. Dye concentrations measured in eluant also are affected by the time and method of elution (Smart and Simpson, 2001), therefore the concentrations of dye measured in eluant have a nonlinear, nonquantifiable relation to the concentrations of dye resurging in water. It is primarily because of these difficulties that some researchers, such as Field (2002b), have expressed reservations about the use of charcoal and strongly advocate the collection and quantitative analysis of water samples, or use of in-situ continuous-flow fluorometery during dye-tracer tests. Assuming that a qualitative tracer test design will meet the objectives for the study, many of these difficulties can be overcome by careful evaluation of ambient fluorescence, careful tracer-test design, proper application of analytical methods, and the application of rigorous QA/QC techniques during all field and laboratory activities. All of these issues deserve careful consideration during the planning phases of a dye-tracer test.

Proper evaluation of ambient fluorescence (background) is even more critical with activated charcoal than with water samples. When used in the field, activated charcoal captures a broad range of organic molecules, and a complex hierarchy of adsorption occurs based on the range of adsorptive sites, their accessibility, and the loading (composition and duration of flow) (Smart and Simpson, 2001). As with tracer dyes, these solutes will be recovered on the charcoal at substantially higher levels than the concentrations present in the water.

During spectrofluorometric analysis, the fluorescent signatures created by these solutes may be confused with, or mask, dye spectral peaks (Smart and others, 1998). High levels of organic solutes can foul the detectors because the solutes can consume the available adsorptive capacity of the carbon. Older charcoal tends to be less adsorptive than fresh charcoal, because of denaturing of the more energetic adsorption sites and capture of organic molecules from the surrounding atmosphere (Smart and Simpson, 2001, 2002).

Unfortunately, there is no ready means of distinguishing a genuine tracer recovery from accidental contamination of the charcoal detector (Smart and others, 1998). Wood charcoals, including coconut shell used to manufacture granular activated charcoal, can contain 10 to 20 percent fluorescein-type functional groups, which may create apparent false-positive peaks for sodium fluorescein dye when eluted (Alexander, 2002); however, this problem is usually manageable in that the compounds generally have a weak fluorescent intensity and seem to be flushed from charcoal by 1 to 2 days exposure to flowing water (Smart and Simpson, 2001).

Dye-Breakthrough Curve Analysis

Analysis of dye-breakthrough curves (measured dye concentration over time) obtained via quantitative dye-tracer tests is an effective means of determining conduit-flow characteristics in karst aquifers (Smoot and others, 1987). Advantages provided by using this method, listed by Kincaid and others (2005), include:

- Plotting of the increase and decrease in fluorescence increases the confidence that tracer-test results are accurate and reflect the actual passage of the injected tracer dye through the aquifer.

- More accurate estimates of flow velocity can be calculated using time-to-peak concentrations.

- Integrating the area under the dye-breakthrough curve allows for estimation of the mass of tracer recovered at a sampling site and, therefore, the relative contribution of flow from the injection site to the tracer resurgence site.

- If it can be assumed that 100 percent of the tracer dye was recovered, evaluation of the shape of the dye-breakthrough curve provides data needed for estimation of hydraulic properties such as longitudinal dispersion, Reynolds and Peclet numbers, and discharge.

Important characteristics of the dye-breakthrough curve (fig. 18) include the first arrival or time to the leading edge of the dye pulse, time to peak concentration, elapsed time of passage of the dye pulse, and time to trailing edge or passage of the dye pulse. As Field (1999a) notes, these characteristics are not entirely objectively defined because they are dependent on sampling frequency and instrument sensitivity. Apart from sampling frequency bias, the shape and magnitude of the dye-breakthrough curve are most influenced by: (1) the amount of dye injected, (2) the velocity and magnitude of the flow, (3) internal structure and hydraulic properties of the conduit flow path taken by the tracer dye, and (4) other factors that affect mixing and dispersion of the tracer dye in the aquifer (Smart, 1998; Field, 1999a). Thus, the dye-breakthrough results obtained represent the transport characteristics of the tracer dye under the hydrologic conditions occurring during a particular test. Repeated quantitative tracer tests may be needed to characterize tracer dye characteristics under different flow conditions. Normalized dye-concentration and dye-load curves are used to compare and evaluate the transport characteristics of dye under different hydrologic conditions (Mull and others, 1988).

The physical properties of the dye-breakthrough curve provide information about conduit structure and organization (Smart, 1998). The dispersion of a dye plume increases with time and distance, and the pattern of dye recovery obtained reflects the effects of processes such as dilution, longitudinal dispersion, divergence, convergence, and storage, which are related to discharge and conduit geometry. The effects of longitudinal dispersion of the dye pulse usually are seen as a lengthening of the breakthrough curve ("tailing"), and the effects of tracer retardation usually are seen as multiple secondary peaks in dye concentration along the profile of the breakthrough curve. Interpretation of complex or multipeaked dye-breakthrough curves may be difficult because the factors contributing to tracer dispersion or retardation may include anastomosing (bifurcation or braiding) conduit-flow paths; flow reversal in eddies and variability in conduit cross-sectional areas (Hauns and others, 2001); intermittent storage and flushing of hydraulically stagnant zones (Smart, 1998); and interconnected zones of higher and lower fracture permeabilities (Shapiro, 2001). The potential effects of such factors on the shapes of dye-breakthrough curves under high-flow and low-flow conditions are illustrated in figure 19. Interpretation of the physical characteristics of the breakthrough curves usually cannot be based solely on the pattern of recovery of dye, but also on knowledge of the physical hydrogeology and conduit structure in the karst aquifer under study (fig. 20) (Jones, 1984b).

A variety of hydraulic properties, including the hydraulic radius or (assuming open-channel flow conditions) hydraulic depth, Peclet number, Reynolds number, Froude number, and hydraulic head loss can be estimated using dye-breakthrough curve data if it can be assumed that nearly 100 percent of the tracer dye was recovered (Field, 1999a; Mull and others, 1988; Field, 2002b). The computer program QTRACER2 (Field, 2002b), automates curve plotting and facilitates many of the calculations involved in the dye-breakthrough curve analysis obtained by analysis of dye-breakthrough curve data.

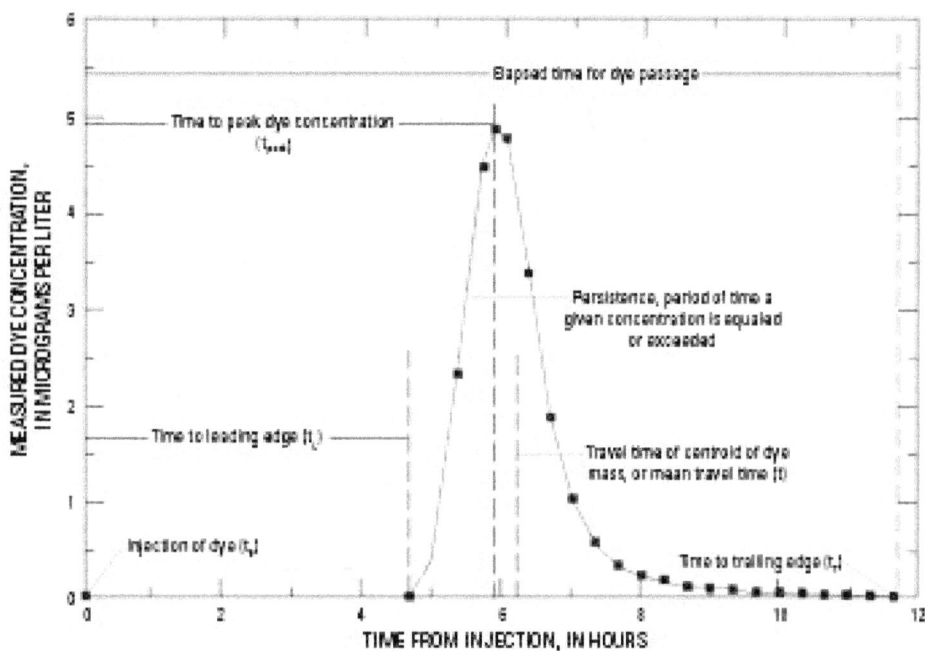

Figure 18.   Some important physical characteristics of a dye-breakthrough curve (from Mull and others, 1988).

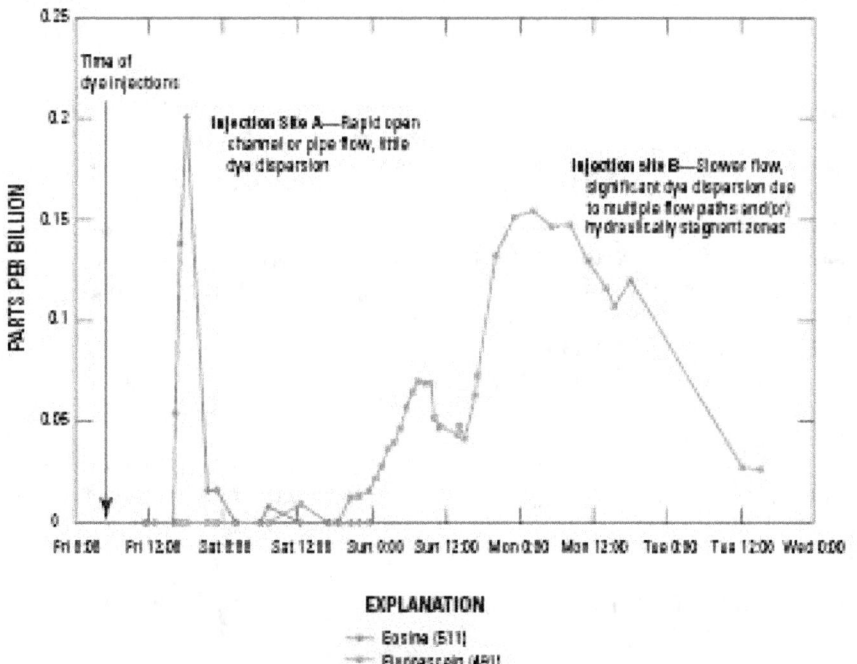

EXPLANATION

— Eosine (511)
— Fluorescein (451)

Figure 19.   Example of dye-breakthrough curves for two dye-tracing tests conducted in the Edwards aquifer, Texas, showing a quick-flow response with little or no dispersion (Injection site A, left), and a slow-flow response showing the effects of dye dispersion (Injection site B, right) (courtesy of Geary Schindel, Edwards Aquifer Authority).

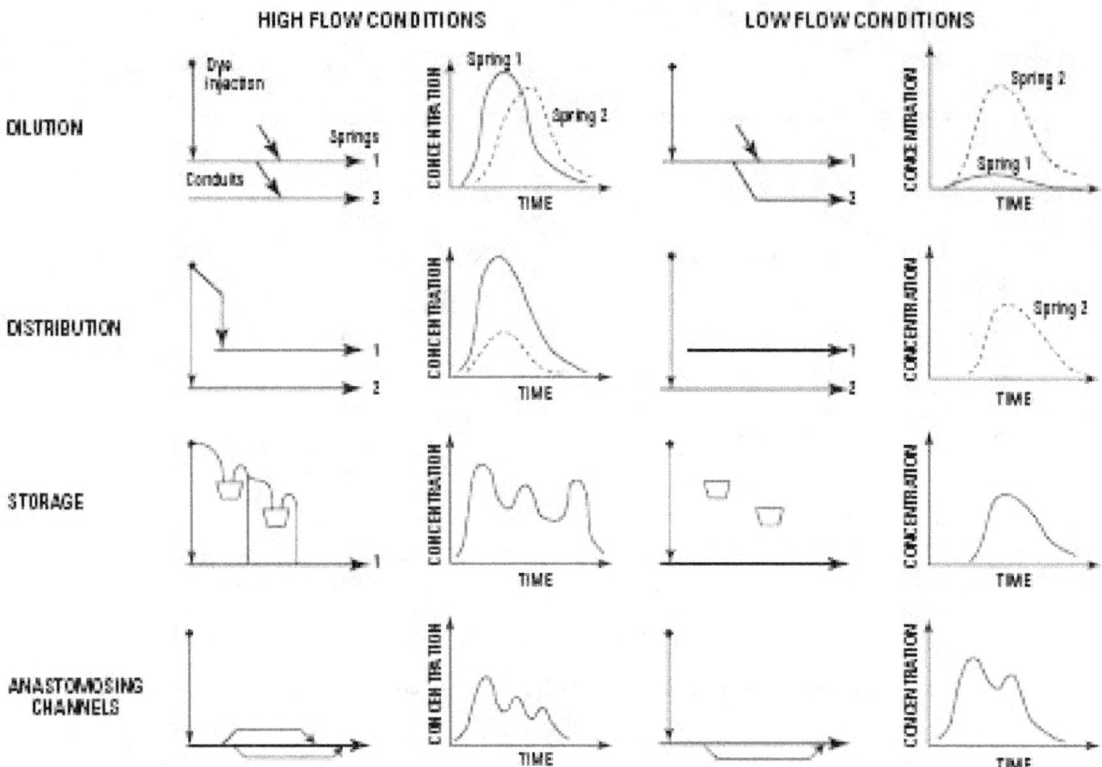

Figure 20.   Shapes of hypothetical dye-breakthrough curves affected by changes in hydrologic conditions (high flow, low flow) and conduit geometry (modified from Jones, 1984b, after Smart and Ford, 1982). Used with permission from the National Speleological Society (www.caves.org).

Mean Tracer-Dye Residence Time

Mean tracer-dye residence time is the length of time required for the centroid (gravity mass) of the tracer dye to traverse the entire length of the karst basin, thus representing the average time of flow through the basin. The centroid generally is not the same as the peak concentration of the tracer-dye mass in the tracer-breakthrough curve, but the more the dye plume conforms to Fick's law (the mass of the diffusing substance passing through a given cross section per unit time is proportional to the concentration gradient) the less obvious the difference between the dye centroid and peak concentration will be.

Mean tracer-dye residence time is estimated by the equation:

$$t_m = \int_0^\infty t\, C(t)\, Q(t)\, dt \,/ \int_0^\infty C(t)\, Q(t)\, dt. \qquad (13)$$

where

t is time of sample collection,

$C(t)$ is measured dye concentration of the sample,

and

$Q(t)$ is the discharge measured at the sampling location.

Tracer-dye residence time will vary from nearly zero for instantaneous transport to almost infinity where the tracer mass is mostly lost to dispersion or storage in the aquifer. If QTRACER2 or another suitable mathematical software program is not used, and the sampling frequency was done at regularly spaced intervals, the integration can be done by using a simple summation algorithm as detailed in Field (2002b) and by Mull and others (1988).

Mean Dye Velocity

Mean tracer velocity (of the dye mass centroid) represents the average rate of travel of dye through the karst basin and is estimated by:

$$V_{(M)} = \int_0^\infty (1.5x\,/\,t)\, C(t)\, Q(t)\, dt \,/ \int_0^\infty C(t)\, Q(t)\, dt. \qquad (14)$$

where

x is straight-line distance between the dye injection and resurgence site.

and

1.5 is a constant representing the conduit sinuosity factor (Field, 1999a).

Tracer Mass Recovery

The accuracy of calculations of mean tracer-dye residence time, flow velocities, and other conduit hydraulic properties from dye-breakthrough curve data is entirely dependent on tracer mass recovery. Few tracing tests result in 100 percent recovery of dye, but as the percentage of mass recovery decreases, the margin of error in the calculated hydraulic parameters increases and confidence in the values obtained declines. Tracer recovery may be affected by the internal structure of conduit networks (Brown and Ford, 1971; Atkinson and others, 1973). It therefore is important to assess tracer mass recovery as a starting point in the analysis of quantitative dye-tracing tests.

The quality of the tracer experiment may be quantified in terms of the relation between the mass of dye tracer injected (M_n) during the experiment and the total mass of dye tracer recovered (M_r). A test accuracy index proposed by Sukhodolov and others (1997) is calculated by:

$$A_I = M_n - M_r / M_n \qquad (15)$$

This index provides a semiquantitative assessment of the quality of the test. A value $A_I = 0$ indicates a perfect tracing experiment with no loss of tracer dye mass. A positive A_I value indicates that more tracer dye mass was injected than was recovered—a common result, whereas a negative value indicates more dye mass was recovered than was injected—an impossibility unless residual tracer dye is present in the aquifer, errors are made in determining the dye concentration in test samples, or initial calculations of the injected dye mass are in error.

In the previous equation, the value for M_r the total mass of tracer dye recovered is given by the equation:

$$M_r = \int_0^\infty C(t)\,Q(t)\,dt. \qquad (16)$$

A simple summation algorithm can be used to facilitate the calculations needed to obtain the value for M_r as described by Field (2002b):

$$M_r = \int_0^\infty C(t)\,Q(t)\,dt \approx , \qquad (17)$$

$$\sum_{t=1}^n C(t)\,Q(t)\,\Delta t_t \approx , \qquad (18)$$

and

$$t_c \sum_{t=1}^n (C_t Q_t), \qquad (19)$$

where

t_c is a time conversion needed to obtain units of mass only.

The previous equations assume that the total dye mass is recovered at a single spring site. If dye has resurged at multiple spring outlets, these calculations are repeated for each site and the results are summed to obtain M_r.

Summary

The hydrogeologic complexities presented by karst terranes often magnify the difficulties involved in identifying and measuring or estimating water fluxes. Conventional hydrogeologic methods such as aquifer tests and potentiometric mapping, though useful, are not completely effective in identifying the processes involved in the transfer of water fluxes in karst, or in characterizing the hydrogeologic framework in which they occur, and may provide erroneous results if data are not collected and interpreted in the context of a karst conceptual model. In karst terranes, a greater emphasis must generally be placed on the identification of hydrologic boundaries and subsurface flow paths, contributions of water from various concentrated and diffuse recharge sources, the hydraulic properties of conduits, and the springs that drain conduit networks. Typically, this emphasis requires the use of a multidisciplinary study approach that includes water-tracer tests conducted with fluorescent dyes and the analysis of spring-discharge and water-chemistry data.

The concepts and methods discussed in this chapter are intended to assist the water-resources investigator in determining what types of data-collection activities may be required for particular karst water-resources management and protection issues, and may aid the planning and implementation of karst hydrogeologic studies. The conceptual model of a karst drainage basin, described herein as a fundamental karst mapping unit defined by the total area of surface and subsurface drainage that contributes water to a conduit network and its outlet spring or springs, may be useful in this regard.

References

Alexander, E.C., Jr., 2002, Scanning spectrofluorophotometer methods: Minneapolis, University of Minnesota Department of Geology and Geophysics (unpublished document dated 5/12/02), 10 p.

Alexander, S.C., 2005, Spectral deconvolution and quantification of natural organic material and fluorescent tracer dyes, in Beck, B.F., ed., Sinkholes and the engineering and environmental impacts of karst, Reston, Virginia: American Society of Civil Engineers, Geotechnical Special Publication No. 144, p. 441–448.

Aley, T., 1977, A model for relating land use and groundwater quality in southern Missouri, in Dilamarter, R.R., and Csallany, S.C., eds., Hydrologic problems in karst regions: Bowling Green, Western Kentucky University, p. 232–332.

Aley, T., 1997, Groundwater tracing in the epikarst, in Beck, B.F., and Stephenson, J.B., eds., The engineering geology and hydrogeology of karst terranes: Rotterdam, Netherlands, A.A. Balkema, p. 207–211.

Arikan, A., 1988, MODALP—A deterministic rainfall-runoff model for large karstic areas: Hydrological Sciences Journal, v. 33, no. 4, p. 401–414.

ASTM Subcommittee D–18–21, 2002, Standard guide for design of ground-water monitoring systems in karst and fractured-rock aquifers: West Conshohocken, Pa., American Society of Testing and Materials, Annual Book of ASTM Standards, v. 04.08, ASTM D5717–95, p. 1421–1438.

Atkinson, T.C., 1977, Diffuse flow and conduit flow in limestone terrain in Mendip Hills, Somerset, England: Journal of Hydrology, v. 35, p. 93–100.

Atkinson, T.C., Smith, D.I., Lavis, J.J., and Whitaker, R.J., 1973, Experiments in tracing underground waters in limestones: Journal of Hydrology, v. 19, p. 323–349.

Baedke, S.J., and Krothe, N.C., 2001, Derivation of effective hydraulic parameters of a karst aquifer from discharge hydrograph analysis: Water Resources Research, v. 37, no. 1, p. 13–19.

Bassett, J., 1976, Hydrology and geochemistry of the upper Lost River drainage basin, Indiana: National Speleological Society Bulletin, v. 38, no. 4, p. 79–87.

Bayless, E.R., Taylor, C.J., and Hopkins, M.S., 1994, Directions of ground-water flow and locations of ground-water divides in the Lost River watershed near Orleans, Indiana: U.S. Geological Survey Water-Resources Investigations Report 94–4195, 25 p., 2 pls.

Beck, B.F., ed., 1995, Karst geohazards: Proceedings of the Fifth Multidisciplinary Conference on Sinkholes and the Engineering and Environmental Impacts of Karst, Gatlinburg, Tenn., April 2–5, 1995: Rotterdam, Netherlands, A.A Balkema, 581 p.

Beck, B.F., 2002, Overview of background resources on karst research emphasizing obscure and "grey area" documents, in Kuniansky, E.L., ed., U.S. Geological Survey Karst Interest Group Proceedings, Shepherdstown, West Virginia, August 20–22, 2002: U.S. Geological Survey Water-Resources Investigations Report 2002–4174, p. 7.

Beck, B.F., ed., 2003, Sinkholes and the engineering and environmental impacts of karst, Proceedings of the Ninth Multidisciplinary Conference, September 6–10, 2003, Huntsville, Alabama: American Society of Civil Engineers, Geotechnical Special Publication no. 122, 737 p.

Beck, B.F., and Stephenson, J.B., eds., 1997, The engineering geology and hydrogeology of karst terranes, Proceedings of the Sixth Multidisciplinary Conference on Sinkholes and the Engineering and Environmental Impacts of Karst, Springfield, Missouri, April 6–9, 1997: Rotterdam, Netherlands, A.A Balkema, 516 p.

Beck, B.F., Pettit, A.J., and Herring, J.G., eds., 1999, Hydrogeology and engineering geology of sinkholes and karst—1999, Proceedings of the Seventh Multidisciplinary Conference on Sinkholes and the Engineering and Environmental Impacts of Karst, Harrisburg-Hershey, Pennsylvania, April 10–14, 1999: Rotterdam, A.A Balkema, 478 p.

Beven, K.J., 2001, Rainfall-runoff modelling—The primer: New York, John Wiley and Sons, Ltd., 360 p.

Beven, K.J., and Kirkby, M.J., 1979, A physically based, variable contributing area model of basin hydrology: Hydrological Science Bulletin, v. 24, no. 1, p. 43–69.

Bonacci, O., 1993, Karst spring hydrographs as indicators of karst aquifers: Hydrological Sciences-Journal-des Sciences Hydrologiques, v. 38, no. 1, p. 51–62.

Bogli, A., 1980, Karst hydrology and physical speleology: New York, Springer-Verlag, 284 p.

Brown, M.C., and Ford, D.C., 1971, Quantitative tracer methods for investigation of karst hydrologic systems: Transactions of the Cave Research Group of Great Britain, v. 13, no. 1, p. 37–51.

Brahana, J.V., and Hollyday, E.F., 1988, Dry stream reaches in carbonate terranes—Surface indicators of ground-water reservoirs: American Water Resources Association Bulletin, v. 24, no. 3, p. 577–580.

Campbell, W.C., Lumsdon-West, M., and Davies, S., 2003, Geographic information system (GIS) support for karst hydrology, in Beck, B.F., ed., Sinkholes and the engineering and environmental impacts of karst, Proceedings of the Ninth Multidisciplinary Conference, September 6–10, 2003, Huntsville, Alabama: American Society of Civil Engineers, Geotechnical Special Publication no. 122, p. 429–438.

Cherkauer, D.S., 2004, Quantifying ground water recharge at multiple scales using PRMS and GIS: Ground Water, v. 42, no. 1, p. 97–110.

Clark, I., and Fritz, P., 1997, Environmental isotopes in hydrogeology: New York, Lewis Publishers, CRC Press LLC, 328 p.

Crawford, N.C., 1987, The karst hydrogeology of the Cumberland Plateau escarpment of Tennessee: Nashville, Tennessee Department of Conservation, Division of Geology, Report of Investigations no. 44, pt. 1, 43 p.

Crawford, N.C., Groves, C.G., Feeney, T.P., and Keller, B.J., 1987, Agriculture and urban nonpoint source pollution impacts on karst aquifers in the Pennyroyal karst region of Kentucky: Bowling Green, Western Kentucky University Center for Cave and Karst Studies (unpublished contract research report prepared for the Division of Water, Kentucky Natural Resources and Environmental Protection Cabinet, Frankfort, Kentucky and the Barren River Area Development District, Bowling Green, Kentucky), 229 p.

Culver, D.C., and White, W.B., 2004, Encyclopedia of caves: Academic Press, Elsevier, 680 p.

Currens, J.C., 1994, Characterization and quantification of nonpoint-source pollutant loads in a conduit-flow-dominated karst aquifer underlying an intensive-use agricultural region, Kentucky: Lexington, Proceedings of the Kentucky Water Resources Symposium, University of Kentucky Water Resources Research Institute, p. 3.

Currens, J.C., and Graham, C.D.R., 1993, Flooding of the Sinking Creek karst area in Jessamine and Woodford Counties, Kentucky: Kentucky Geological Survey Report of Investigations 7, series XI, 33 p.

Currens, J.C., and Ray, J.A., 1999, Karst atlas for Kentucky, in Beck, B.F., Pettit, A.J., and Herring, J.G., eds., Hydrogeology and engineering geology of sinkholes and karst, Proceedings of the Seventh Multidisciplinary Conference on Sinkholes and the Engineering and Environmental Impacts of Karst, April 10–14, 1999, Harrisburg-Hershey, Pennsylvania: Rotterdam, Netherlands, A.A. Balkema, p. 85–90.

Currens, J.C., 2001, Generalized block diagram of the Western Pennyroyal Karst: Kentucky Geological Survey, map and chart 16, series XII.

Doctor, D.H., and Alexander, E.C., Jr., 2005, Interpretation of water chemistry and stable isotope data from a karst aquifer according to flow regimes identified through hydrograph recession analysis, in Kuniansky, E.L., ed., U.S. Geological Survey Karst Interest Group Proceedings, Rapid City, South Dakota, September 12–15, 2005: U.S. Geological Survey Scientific Investigations Report 2005–5160, p. 82–92.

Dogwiler, T., and Wicks, C.M., 2004, Sediment entrainment and transport in fluviokarst systems: Journal of Hydrology, v. 295, p. 163–172.

Dooge, J.C.I., 1973, Linear theory of hydrologic systems: U.S. Department of Agriculture Technical Bulletin 1468, 327 p.

Dreiss, S.J, 1982, Linear kernels for karst aquifers: Water Resources Research, v. 18, no. 4, p. 865–876.

Dreiss, S.J, 1983, Linear unit-response functions as indicators of recharge areas for karst springs: Journal of Hydrology, v. 61, p. 31–44.

Dreiss, S.J, 1989, Regional scale transport in a karst aquifer, 2—Linear systems and time moment analysis: Water Resources Research, v. 25, no. 1, p. 126–134.

Duley, J.W., 1986, Water tracing using a scanning spectrofluorometer for detection of fluorescent dyes, in Environmental Problems in Karst Terranes and their Solutions Conference, Bowling Green, Kentucky: Dublin, Ohio, Proceedings of the National Water Well Association, October 28–30, 1986, p. 389–406.

Felton, G.K., and Currens, J.C., 1994, Peak flow rate and recession-curve characteristics of a karst spring in the Inner Bluegrass, Central Kentucky: Journal of Hydrology, v. 162, p. 99–118.

Field, M.S., 1999a, Quantitative analysis of tracer breakthrough curves from tracing tests in karst aquifers, in Palmer, A.N., Palmer, M.V., and Sasowsky, I.D., eds., Karst modeling: Leesburg, Va., Karst Waters Institute Special Publication 5, p. 163–171.

Field, M.S., 1999b, On the importance of stock dye concentrations for accurate preparation of calibration standards, in Palmer, A.N., Palmer, M.V. and Sasowsky, I.D., eds., Karst modeling: Leesburg, Va., Karst Waters Institute, Special Publication 5, p. 229.

Field, M.S., 2002a, A lexicon of cave and karst terminology with special reference to environmental karst hydrology: United States Environmental Protection Agency Publication EPA/600/R-02/003, 214 p.

Field, M.S., 2002b, The QTRACER2 program for tracer-breakthrough curve analysis for tracer tests in karstic aquifers and other hydrologic systems: U.S. Environmental Protection Agency Publication EPA/600/R-02/001, 179 p.

Field, M.S., 2003, Tracer-test planning using the Efficient Hydrologic Tracer-Test Design (EHTD) Program: Washington, D.C., U.S. Environmental Protection Agency, Office of Research and Development, National Center for Environmental Assessment, EPA/600/R-03/034, 103 p.

Field, M.S., 2006, Tracer-test design for losing stream-aquifer systems: International Journal of Speleology, v. 35, no. 1, p. 25–36.

Field, M.S., and Nash, S.G., 1997, Risk assessment methodology for karst aquifers—(1) Estimating karst conduit-flow parameters: Environmental Monitoring and Assessment, v. 47, no. 1, p. 1–27.

Field, M.S., Wilhelm, R.G., Quinlan, J.F., and Aley, T.J., 1995, An assessment of the potential adverse properties of fluorescent tracer dyes used for groundwater tracing: Environmental Monitoring and Assessment, v. 38, no. 1, p. 75–96.

Ford, D.C., 1999, Perspectives in karst hydrogeology and cavern genesis, in Palmer, A.N., Palmer, M.V., and Sasowsky, I.D., eds., Karst modeling: Leesburg, Va., Karst Waters Institute Special Publication 5, p. 17–29.

Ford, Derek, and Williams, P.W., 1989, Karst geomorphology and hydrology: London, Unwin Hyman, 601 p.

George, A.I., 1989, Caves and drainage north of the Green River in White, W.B., and White, E.L., eds., Karst hydrology concepts from the Mammoth Cave area: New York, Van Nostrand Reinhold, p. 189–222.

Ginsberg, M., and Palmer, A., 2002, Delineation of source-water protection areas in karst aquifers of the Ridge and Valley and Appalachian Plateaus Physiographic Provinces—Rules of thumb for estimating the capture zones of springs and wells: U.S. Environmental Protection Agency EPA 816–R–02–015, 41 p.

Greene, E.A., 1997, Tracing recharge from sinking streams over spatial dimensions of kilometers in a karst aquifer: Ground Water, v. 35, no. 5, p. 898–904.

Greene, E.A., Shapiro, A.M., and Carter, J.M., 1999, Hydrogeologic characterization of the Minnelusa and Madison aquifers near Spearfish, South Dakota: U.S. Geological Survey Water-Resources Investigations Report 98–4156, 64 p.

Gunn, J., 1983, Point-recharge of limestone aquifers—A model from New Zealand karst: Journal of Hydrology, v. 61, p. 19–29.

Gunn, J., 1986, A conceptual model for conduit flow dominated karst aquifers, in Gunay, G., and Johnson, A.I., eds., Karst water resources, Proceedings of the International Symposium, Ankara, Turkey, July 1985: International Association of Hydrological Sciences Publication 161, p. 587–596.

Hauns, M., Jeannin, P.-Y., and Atteia, O., 2001, Dispersion, retardation and scale effect in tracer breakthrough curves in karst conduits: Journal of Hydrology, v. 241, p. 177–193.

Hess, J.W., and White, W.B., 1989, Water budget and physical hydrology, in White, W.B., and White, E.L., eds., Karst hydrology concepts from the Mammoth Cave area: New York, Van Nostrand Reinhold, p. 105–126.

Howcroft, W.D., 1992, Ground water flow and water resources investigation of the Auburn, Summers, and Shakertown Springs karst groundwater basins, Logan and Simpson Counties, Kentucky: Bowling Green, Western Kentucky University, Department of Geography, unpublished thesis, 128 p.

Imes, J.L., Schumacher, J.G., and Kleeschulte, M.J., 1996, Geohydrologic and water-quality assessment of the Fort Leonard Wood Military Reservation, Missouri, 1994–95: U.S. Geological Survey Water-Resources Investigations Report 96–4270, 134 p.

Jones, W.K., 1984a, Dye tracer tests in karst areas: National Speleological Society Bulletin, v. 46, p. 3–9.

Jones, W.K., 1984b, Analysis and interpretation of data from tracer tests in karst areas: National Speleological Society Bulletin, v. 46, p. 41–47.

Jones, W.K., 1997, Karst hydrology atlas of West Virginia: Leesburg, Va., Karst Waters Institute Special Publication 4, 111 p.

Jones, W.K., Culver, D.C., and Herman, J.S., eds., 2004, Epikarst, Proceedings of the 2003 Epikarst symposium, October 1–4, 2003, Shepherdstown, West Virginia, USA: Leesburg, Va., Karst Waters Institute Special Publication 9, 160 p.

Kasnavia, T., Vu, D., and Sabatini, D.A., 1999, Fluorescent dye and media properties affecting sorption and tracer selection: Ground Water, v. 37, no. 3, p. 376–381.

Kass, W., 1998, Tracing technique in geohydrology: Rotterdam, Netherlands, A.A. Balkema; Brookfield, Vermont, A.A. Balkema Publishers, 581 p.

Katz, B.G., 2005, Demystification of ground-water flow and contaminant movement in karst systems using chemical and isotopic tracers, in Kuniansky, E.L., ed., U.S. Geological Survey Karst Interest Group Proceedings, Shepherdstown, West Virginia August 20–22, 2002: U.S. Geological Survey Water-Resources Investigations Report 2002–4174, p. 13–19.

Katz, B.G., Coplen, T.B., Bullen, T.D., and Davis, J.H., 1997, Use of chemical and isotopic tracers to characterize the interactions between ground water and surface water in mantled karst: Ground Water, v. 35, no. 6, p. 1014–1028.

Kilpatrick, F.A., and Cobb, E.D., 1984, Measurement of discharge using tracers: U.S. Geological Survey Open-File Report 84–136, 73 p.

Kincaid, T.R., Hazlett, T.J., and Davies, G.J., 2005, Quantitative groundwater tracing and effective numerical modeling in karst—An example from the Woodville Karst Plain of North Florida, in Beck, B.F., ed., Sinkholes and the engineering and environmental impacts of karst, Proceedings of the 10th Multidisciplinary Conference, San Antonio, Texas, September 24–28, 2005: Geological Institute of the American Society of Civil Engineers, Geotechnical Special Publication no. 144, p. 114–121.

Klimchouk, A.B., 2004, Towards defining, delimiting and classifying epikarst—Its origin, processes and variants of geomorphic evolution, in Jones, W.K., Culver, D.C., and Herman, J., eds., Epikarst, Proceedings of the 2003 Epikarst symposium, October 1–4, 2003, Shepherdstown, West Virginia, USA: Karst Waters Institute Special Publication 9, p. 23–35.

Klimchouk, A.B., Ford, D.C., Palmer, A.N., and Dreybrodt, W., 2000, Speleogenesis evolution of karst aquifers: National Speleological Society, 527 p.

Kresic, N., 1997, Quantitative solutions in hydrogeology and groundwater modeling: Boca Raton, CRC Lewis Publishers, 461 p.

Kuniansky, E.L., ed., 2001, U.S. Geological Survey Karst Interest Group Proceedings, St. Petersburg, Florida, February 13–16, 2001: U.S. Geological Survey Water-Resources Investigations Report 2001–4011, 211 p.

Kuniansky, E.L., ed., 2002, U.S.Geological Survey Karst Interest Group Proceedings, Shepherdstown, West Virginia, August 20–22, 2002: U.S. Geological Survey Water-Resources Investigations Report 2002–4174, 89 p.

Kuniansky, E.L. ed., 2005, U.S.Geological Survey Karst Interest Group Proceedings, Rapid City, South Dakota, September 12–15, 2005: U.S. Geological Survey Scientific Investigations Report 2005–5160, 296 p.

Kuniansky, E.L., Fahlquist, L., and Ardis, A.F., 2001, Travel times along selected flow paths of the Edwards aquifer, Central Texas, in Kuniansky, E.L., ed., U.S. Geological Survey Karst Interest Group Proceedings, St. Petersburg, Florida, February 13–16, 2001: U.S. Geological Survey Water-Resources Investigations Report 2001–4011, p. 69–77.

Lakey, B.L., and Krothe, N.C., 1996, Stable isotope variation of storm discharge from a perennial karst spring, Indiana: Water Resources Research, v. 32, no. 2, p. 721–731.

Lane, S.R., and Smart, C.C., 1999, A method for correction of variable background fluorescence in filter fluorometry, in Beck, B.F., Pettit, A.J., and Herring, J.G., eds., Hydrology and engineering geology of sinkholes and karst—1999, Proceedings of the Seventh Multidisciplinary Conference, Harrisburg-Hershey Pennsylvania, April 10–14, 1999: Rotterdam, Netherlands, A.A Balkema, p. 301–306.

Leavesley, G.H., Lichty, R.W., Troutman, B.M., and Saindon, L.G., 1983, Precipitation-Runoff Modeling System—User's manual: U.S. Geological Survey Water-Resources Investigations Report 83–4238, 207 p.

Lee, E.S., and Krothe, N.C., 2001, A four-component mixing model for water in a karst terrain in south-central Indiana, USA, using solute concentration and stable isotopes as tracers: Chemical Geology, v. 179, p. 129–143.

Liedl, R., Sauter, M., Hückinghaus, D., Clemens, T., and Teutsch, G., 2003, Simulation of the development of karst aquifers using a coupled continuum pipe flow model: Water Resources Research, v. 39, no. 3, p. 1057.

Long, A.J., and Derickson, R.G., 1999, Linear systems analysis in a karst aquifer: Journal of Hydrology, v. 219, p. 206–217.

Meiman, J., and Ryan, M.T., 1999, The development of basin-scale conceptual models of the active-flow conduit system, in Palmer, A.N., Palmer, M.V., and Sasowsky, I.D., eds., Karst modeling: Leesburg, Va., Karst Waters Institute Special Publication 5, p. 203–212.

Meinzer, O.E., 1927, Large springs in the United States: U.S. Geological Survey Water-Supply Paper 557, 94 p.

Milanović, P.T., 1981a, Water regime in deep karst—Case study of the Ombla Spring drainage area, in Yevjevich, V., ed., Karst hydrology and water resources, Proceedings of the U.S.-Yugoslavian Symposium, Dubrovnik, 1975: Littleton, Colo., Water Resources Publication, p. 165–191.

Milanović, P.T., 1981b, Karst hydrogeology: Littleton, Colo., Water Resources Publication, 434 p.

Mull, D.S., Smoot, J.L., and Liebermann, T.D., 1987, Dye tracing techniques used to determine ground-water flow in a carbonate aquifer system near Elizabethtown, Kentucky: U.S. Geological Survey Water-Resources Investigations Report 87–4174, 95 p., 2 pls.

Mull, D.S., Liebermann, T.D., Smoot, J.L., and Woosley, Jr., L.H., 1988, Application of dye-tracing techniques for determining solute-transport characteristics of ground water in karst terranes: Atlanta, Ga., U.S. Environmental Protection Agency, EPA Report 904/6–88–001, 103 p.

National Academy of Sciences, 2004, Groundwater fluxes across interfaces—Committee on hydrologic science: Washington, D.C., National Research Council, National Academies Press, 100 p.

National Ground Water Association, 1991, Proceedings of the Third Conference on Hydrogeology, Ecology, Monitoring, and Management of Ground Water in Karst Terranes, December 4–6, 1991, Nashville, Tennessee: Dublin, Ohio, National Ground Water Association, 793 p.

National Water Well Association, 1986, Proceedings of the Environmental Problems in Karst Terranes and Their Solutions Conference, October 28–30, 1986, Bowling Green, Kentucky: Dublin, Ohio, National Water Well Association, 525 p.

National Water Well Association, 1988, The Proceedings of the Second Conference on Environmental Problems in Karst Terranes and Their Solutions, November 16–18, 1988, Nashville, Tennessee: Dublin, Ohio, National Water Well Association, 441 p.

Neuman, S.P., and de Marsily, G., 1976, Identification of linear systems response by parametric programming: Water Resources Research, v. 12, no. 2, p. 253–262.

Padilla, A., Pulio-Bosch, A., and Mangin, A., 1994, Relative importance of baseflow and quickflow from hydrographs of karst springs: Ground Water, v. 32, p. 267–277.

Paillet, F.L., 2001, Borehole geophysical applications in karst hydrology, in Kuniansky, E.L., ed., U.S. Geological Survey Karst Interest Group Proceedings, St. Petersburg, Florida, February 13–16, 2001: U.S. Geological Survey Water-Resources Investigations Report 2001–4011, p. 116–123.

Palmer, A.N., 1991, Origin and morphology of limestone caves: Geological Society of America Bulletin, v. 103, p. 1–21.

Palmer, A.N., 1999, Patterns of dissolution porosity in carbonate rocks, in Palmer, A.N., Palmer, M.V., and Sasowsky, I.D., eds., Karst modeling: Leesburg, Va., Karst Waters Institute Special Publication 5, p. 71–78.

Pinault, J.L., Plagnes, V., Aquilina, L., and Bakalowicz, M., 2001, Inverse modeling of the hydrological and hydrochemical behavior of hydrosystems—Characterization of karst system functioning: Water Resources Research, v. 37, no. 8, p. 2191–2204.

Quinlan, J.F., 1970, Central Kentucky karst: Méditerranée Études et Travaux, v. 7, p. 235–253.

Quinlan, J.F., 1989, Ground-water monitoring in karst terranes—Recommended protocols and implicit assumptions: Las Vegas, Nev., U.S. Environmental Protection Agency, Environmental Monitoring Systems Laboratory, EPA/600/X–89/050, 79 p.

Quinlan, J.F., and Ewers, R.O., 1989, Subsurface drainage in the Mammoth Cave area, in White, W.B., and White, E.L., eds., Karst hydrology concepts from the Mammoth Cave area: New York, Van Nostrand Reinhold, p. 65–104.

Quinlan, J.F., Ewers, R.O., Ray, J.A., Powell, R.L., and Krothe, N.C., 1983, Groundwater hydrology and geomorphology of the Mammoth Cave region, Kentucky, and of the Mitchell Plain, Indiana, in Shaver, R.H., and Sunderman, V.A., eds., Field trips in midwestern geology: Geological Society of America, v. 2, p. 1–85.

Quinlan, J.F., and Ray, J.A., 1989, Groundwater basins in the Mammoth Cave region, Kentucky, showing springs, major caves, flow routes, and potentiometric surface: Mammoth Cave, Ky., Friends of the Karst Occasional Publication no. 2, 1 pl.

Quinlan, J.F., and Ray J.A., 1995, Normalized base-flow discharge of ground-water basins—A useful parameter for estimating recharge area of springs and for recognizing drainage anomalies in karst terranes, in Beck, B., ed., Karst geohazards, Proceedings of the Fifth Multidisciplinary Conference on Sinkholes and the Engineering and Environmental Impacts of Karst, April 2–5, 1995, Gatlinburg, Tennessee: Rotterdam, Netherlands, A.A. Balkema, p. 149–164.

Quinlan, J.F., Ray, J.A., and Schindel, G.M., 1995, Intrinsic limitations of standard criteria and methods for delineation of groundwater-source protection areas (springhead and wellhead protection areas) in carbonate terranes—Critical review, technically-sound resolution of limitations, and case study in a Kentucky karst, in Beck, B.F., and Pearson, F.M., eds., Karst geohazards engineering and environmental problems in karst terrane: A.A. Balkema, Rotterdam, Netherlands, p. 165–176.

Ray, J.A., 1994, Surface and subsurface trunk flow, Mammoth Cave, Kentucky: Mammoth Cave, Ky., Proceedings of the Third Mammoth Cave National Park Science Conference, Mammoth Cave National Park, p. 175–187.

Ray, J., 1999, A model of karst drainage basin evolution, interior low plateaus, USA, in Palmer, A.N., Palmer, M.V., and Sasowsky, I.D., eds., Karst modeling: Leesburg, Va., Karst Waters Institute Special Publication 5, p. 58.

Ray, J., 2001, Spatial interpretation of karst drainage basins: in Beck, B.F., and Herring, J.G., eds., Geotechnical and environmental applications of karst geology and hydrology, Proceedings of the Eighth Multidisciplinary Conference on Sinkholes and the Environmental and Engineering Impacts of Karst, April 1–4, 2001, Louisville, Kentucky: Lisse, Balkema Publishers, p. 235–244.

Ray, J.A., and Currens, J.C., 1998, Mapped karst ground-water basins in the Beaver Dam 30 × 60 minute quadrangle: Kentucky Geological Survey Map and Chart, ser. 11, map and chart 19, scale 1:100,000.

Rendell, D., 1987, Fluorescence and phosphorescence—Analytical chemistry by open learning series: London, Thames Polytechnic, John Wiley and Sons, 419 p.

Rorabaugh, M.I., 1964, Estimating changes in bank storage as ground-water contribution to streamflow: International Association Scientific Hydrological Publications 63, p. 432–441.

Rovey, C.W., II, and Cherkauer, D.S., 1994, Relation between hydraulic conductivity and texture in a carbonate aquifer: regional continuity: Groundwater v. 32, no. 2, p. 227–238.

Ryan, M.T., and Meiman, J., 1996, An examination of short-term variations in water quality at a karst spring in Kentucky: Ground Water, v. 34, no. 1, p. 23–30.

Sauter, M., 1991, Assessment of hydraulic conductivity in a karst aquifer at local and regional scale, in Proceedings of the Third Conference on Hydrogeology, Ecology, Monitoring, and Management of Ground Water in Karst Terranes December 4–6, 1991, Nashville, Tennessee: National Ground Water Association, p. 39–56.

Sauter, M., 1992, Assessment of hydraulic conductivity in a karst aquifer at local and regional scale: Ground Water Management, v. 10, p. 39–57.

Scanlon, B.R., Mace, R.E., Barrett, M.E., and Smith, B., 2003, Can we simulate regional groundwater flow using equivalent porous media models? Case study, Barton Springs Edwards aquifer, USA: Journal of Hydrology, v. 276, p. 137–158.

Scanlon, B.R., and Thrailkill, 1987, Chemical similarities among physically distinct spring types in a karst terrain: Journal of Hydrology, v. 89, p. 259–279.

Schindel, G.M., Ray, J.A., and Quinlan, J.F., 1995, Delineation of the recharge area for Rio Springs, Kentucky—An EPA demonstration project in wellhead (springhead) protection for karst terranes, *in* Beck, B.F., and Pearson, F.M., eds., Karst geohazards engineering and environmental problems in karst terrane: Rotterdam, Netherlands, and Brookfield, Vermont, A.A. Balkema, p. 165–176.

Shapiro, A.M., 2001, Effective matrix diffusion in kilometer-scale transport in fractured crystalline rock: Water Resources Research, v. 37, p. 507–522.

Shevenell, L., 1996, Analysis of well hydrographs in a karst aquifer—Estimates of specific yields and continuum transmissivities: Journal of Hydrology, v. 174, p. 331–355.

Shuster, E.T., and White, W.B., 1971, Seasonal fluctuations in the chemistry of limestone springs—A possible means of characterizing carbonate aquifers: Journal of Hydrology, v. 14, p. 93–128.

Smart, C.C., 1988, A deductive model of karst evolution based on hydrological probability: Earth Surface Processes and Landforms, v. 13, p. 271–288.

Smart, C.C., 1998, Artificial tracer techniques for the determination of the structure of conduit aquifers: Ground Water, v. 26, p. 445–453.

Smart, C.C., 2005, Error and technique in fluorescent dye tracing, *in* Beck, B.F., ed., Sinkholes and the engineering and environmental impacts of karst: Reston, Va., American Society of Civil Engineers, Geotechnical Special Publication no. 144, p. 500–509.

Smart, C.C., and Ford, D.C., 1982, Quantitative dye tracing in a glacierized alpine karst: Beitraege Zur Geologie der Schweiz—Hydrologie, v. 28, no. 1, p. 191–200.

Smart, C.C., and Karunaratne, K.C., 2002, Characterization of fluorescence background in dye tracing: Environmental Geology, v. 42, p. 492–498.

Smart, C.C., and Simpson, B., 2001, An evaluation of the performance of activated charcoal in detection of fluorescent compounds in the environment, *in* Beck and Herring, eds., Geotechnical and environmental applications of karst geology and hydrology: Rotterdam, Netherlands, A.A Balkema, p. 265–270.

Smart, C.C., and Simpson, B., 2002, Detection of fluorescent compounds in the environment using granular activated charcoal detectors: Environmental Geology, v. 42, p. 538–545.

Smart, C.C., and Smart, P.L., 1991, Correction of background interference and cross-fluorescence in filter fluorometric analysis of water-tracer dyes, *in* Proceedings of the Third Conference on Hydrogeology, Ecology, Monitoring, and Management of Ground Water in Karst Terranes, December 4–6, 1991, Nashville, Tennessee: National Ground Water Association, p. 475–491.

Smart, C.C., Zabo, L., Alexander, E.C., Jr., and Worthington, S.R.H., 1998, Some advances in fluorometric techniques for water tracing: Environmental Monitoring and Assessment, v. 53, p. 305–320.

Smart, P.L., 1972, A laboratory evaluation of the use of activated carbon for the detection of tracer dye Rhodamine WT: University of Alberta, Canada, (unpublished Master's thesis).

Smart, P.L., and Laidlaw, I.M.S., 1977, An evaluation of some fluorescent dyes for water tracing: Water Resources Research, v. 13, p. 15–33.

Smoot, J.L., Mull, D.S., and Liebermann, T.D., 1987, Quantitative dye tracing techniques for describing the solute transport characteristics of ground-water flow in karst terrane, *in* Proceedings of the Second Multidisciplinary Conference on Sinkholes and the Environmental Impacts of Karst, Orlando, Florida, February 9–11, 1987: National Water Well Association, p. 269–275.

Steinheimer, T.R., and Johnson, S.M., 1986, Investigation of the possible formation of diethylnitrosamine resulting from the use of rhodamine WT as a tracer in river waters: U.S. Geological Survey Water-Supply Paper 2290, p. 37–49.

Stoica, P., and Soderstrom, T., 1982, Instrumental variable methods for identification of Hammerstein system: International Journal of Control, v. 35, p. 459–476.

Streltsova, T.D., 1988, Well testing in heterogeneous formations: New York, John Wiley and Sons, 413 p.

Sukhodolov, A.N., Nikora, V.I., Rowiński, P.M., and Czernuszenko, W., 1997, A case study of longitudinal dispersion in small lowland rivers: Water Environmental Resources, v. 97, p. 1246–1253.

Taylor, C.J., and McCombs, G.K., 1998, Recharge-area delineation and hydrology, McCraken Springs, Fort Knox Military Reservation, Meade County, Kentucky: U.S. Geological Survey Water-Resources Investigations Report 98–4196, 12 p., 1 pl.

Taylor, C.J., Nelson, H.L., Hileman, G., and Kaiser, W.P., 2005, Hydrogeologic framework mapping of shallow, conduit-dominated karst—Components of a regional GIS-based approach, *in* Kuniansky, E.L., ed., U.S. Geological Survey Karst Interest Group Proceedings, Rapid City, South Dakota, September 12–15, 2005: U.S. Geological Survey Scientific Investigations Report 2005–5160, p. 103–113.

Teutsch, G., 1992, Groundwater modeling in karst terranes—Scale effects, data acquisition, and field validation: Ground Water Management, v. 10, p. 17–35.

Teutsch G., and Sauter, M., 1991, Groundwater modeling in karst terranes—Scale effects, data acquisition, and field verification, Proceedings of the Third Conference on Hydrogeology, Ecology, Monitoring, and Management of Ground Water in Karst Terranes, Nashville, Tennessee, December 4–6, 1991: National Ground Water Association, p. 17–54.

Thrailkill, J., 1985, The Inner Bluegrass karst region, in Dougherty, P.H., ed., Caves and karst of Kentucky: Kentucky Geological Survey Special Publication 12, Series IX, p. 28–62.

Toth, J., 1963, A theoretical analysis of groundwater flow in small drainage basins: Journal of Geophysical Research, v. 68, p. 211–222.

Trček, B., and Krothe, N.C., 2002, The importance of three and four-component storm hydrograph separation techniques for karst aquifers, in Gabrovšek, Franci, ed., Evolution of karst—From prekarst to cessation: Postojna-Ljubljana Založba ZRC, p. 395–401.

Tucker, R.B., and Crawford, N.C., 1999, Non-linear curve-fitting analysis as a tool for identifying and quantifying multiple fluorescent tracer dyes present in samples analyzed with a spectrofluorophotometer and collected as part of a dye tracer study of groundwater flow, in Beck, B.F., Pettit, A.J., and Herring, J.G., eds., Hydrology and engineering geology of sinkholes and karst—1999: Harrisburg-Hershey, Pennsylvania, Proceedings of the Seventh Multidisciplinary Conference, April 10–14, 1999, p. 307–312.

U.S. Geological Survey, 2005, User's Manual for the National Water Information System of the U.S. Geological Survey, Ground-Water Site-Inventory System: U.S. Geological Survey Open-File Report, version 4.5, available at <http://www.nwis.er.usgs.gov/nwisdocs4_5/gw/gw_complete.pdf>

Vandike, J.E., 1992, The hydrogeology of the Bennett Spring area, Laclede, Dallas, Webster, and Wright Counties, Missouri: Missouri Department of Natural Resources Division of Geology and Land Survey Water Resources Report no. 38, 112 p.

White, W.B. 1988, Geomorphology and hydrology of karst terrains: New York, Oxford University Press, 464 p.

White, W.B., 1993, Analysis of karst aquifers, in Alley, W.M., ed., Regional ground-water quality: New York, Van Nostrand Reinhold, p. 471–489.

White, W.B., 1999, Conceptual models for karstic aquifers, in Palmer, A.N., Palmer, M.V., and Sasowsky, I.D., eds., Karst modeling: Leesburg, Va., Karst Waters Institute Special Publication 5, p. 11–16.

White, W.B., and Schmidt, V.A., 1966, Hydrology of a karst area in east-central West Virginia: Water Resources Research, v. 2, no. 3, p. 549–560.

White, W.B., and White, E.L., eds., 1989, Karst hydrology—Concepts from the Mammoth Cave area: New York, Van Nostrand Reinhold, 346 p.

Wicks, C.M., and Hoke, J.A., 2000, Prediction of the quantity and quality of Maramec Spring water: Ground Water, v. 38, no. 2, p. 218–225.

Williams, P.W, 1983, The role of the subcutaneous zone in karst hydrology: Journal of Hydrology, v. 61, p. 45–67.

Wilson, J.F., Jr., Cobb, E.D., and Kilpatrick, F.A., 1986, Fluorometeric procedures for dye tracing: U.S. Geological Survey Techniques for Water Resources Investigations, book 3, chap. A12, 34 p.

Worthington, S.R.H., 1991, Karst hydrology of the Canadian Rocky Mountains: Hamilton, Ontario, Canada, McMaster University, unpublished Ph. D. dissertation, 380 p.

Worthington, S.R.H., 1999, A comprehensive strategy for understanding flow in carbonate aquifers, in Palmer, A.N., Palmer, M.V., and Sasowsky, I.D., eds., Karst modeling: Leesburg, Va., Karst Waters Institute Special Publication 5, p. 30–37.

Worthington, S.R.H., 2003, Numerical simulation of the aquifer at Mammoth Cave, in Kambesis, P., Meiman, J., and Groves, C., eds., Proceedings of the International Conference on Karst Hydrogeology and Ecosystems, Bowling Green, Kentucky, USA, June 3–6, 2003: Western Kentucky University, Hoffman Environmental Research Institute, p. 10.

Worthington, S.R.H., Davies, G.J., and Ford, D.C., 2000, Matrix, fracture and channel components of storage and flow in a Paleozoic limestone aquifer, in Sasowsky, I.D., and Wicks, C.M., eds., Groundwater flow and contaminant transport in carbonate aquifers: Rotterdam, Netherlands, A.A. Balkema, p. 113–128.

Worthington, S.R.H., and Smart, C.C., 2003, Empirical determination of tracer mass for sink to spring tests in karst, in Beck, B., ed., Sinkholes and the engineering and environmental impacts of karst: Huntsville, Ala., Proceedings of the Ninth Multidisciplinary Conference, September 6–10, 2003, p. 287–295.

Zhang, Y.K., and Bai, E.W., 1996, Simulation of spring discharge from a limestone aquifer in northeastern Iowa with an identification scheme, in Kovar, Karl, and van der Heijde, Paul, eds., Calibration and reliability in groundwater modeling: Golden, Colo., Proceedings of ModelCARE96 Conference, September 1996, IAHS Publication no. 237, p. 31–40.

Zhang, Y.K., Bai, E.W., Rowden, R., and Liu, H., 1995, Simulation of spring discharge from a limestone aquifer in Iowa, USA: Hydrogeology Journal, v. 4, no. 4, p. 41–54.

Analysis of Temperature Gradients to Determine Stream Exchanges with Ground Water

By James E. Constantz, Richard G. Niswonger, and Amy E. Stewart

Chapter 4 of
Field Techniques for Estimating Water Fluxes Between Surface Water and Ground Water

Edited by Donald O. Rosenberry and James W. LaBaugh

Techniques and Methods Chapter 4–D2

U.S. Department of the Interior
U.S. Geological Survey

Contents

Figures

Chapter 4
Analysis of Temperature Gradients to Determine Stream Exchanges with Ground Water

By James E. Constantz,[1] Richard G. Niswonger,[2] and Amy E. Stewart[3]

Introduction

Heat flows continuously between surface water and adjacent ground water, and as a consequence, provides an opportunity to use heat as a natural tracer of water movement between the surface and the underlying sediments. By the early 1900s, researchers intuitively understood that heat is simultaneously transferred during the course of water movement through sediments and other porous bodies (Bouyoucos, 1915). Examination of temperature patterns provided qualitative and quantitative descriptions of an array of ground-water-flow regimes, ranging from those beneath rice paddies to those beneath volcanoes. Quantitative analysis of heat and water flow was introduced via analytical and numerical solutions to the governing partial differential equations. These quantitative analyses often relied on field measurements for parameter identification and accurate predictions of flow rates and directions. Because field measurements of temperature had to be made manually, however, the data were sparse. Early numerical simulation of heat and mass ground-water transport required significant computational resources, which limited modeling to conceptual demonstrations. As a consequence of these challenges, the use of heat as a tracer of ground-water movement was confined to isolated research projects, which could demonstrate only the feasibility of the method rather than progressing toward a routine use of the technique. Recently, both the measurement of temperature and the simulation of heat and water transport have benefited from significant advances in data acquisition and computer resources. This has afforded the opportunity for routine use of heat as a tracer in a variety of hydrological regimes. The measurement of heat flow is particularly well suited for investigations of stream/ground-water exchanges. Dynamic temperature patterns between a stream and the underlying sediments are typical, because of large stream surface area to volume ratios relative to many other surface-water bodies. Heat is a naturally occurring tracer, free from (real or perceived) issues of

contamination associated with the use of chemical tracers in stream environments. The use of heat as a tracer relies on the measurement of temperature gradients, and temperature is an extremely robust property to monitor. Temperature data are immediately available as opposed to most chemical tracers, many of which require laboratory analysis. The recent publication of numerous case studies (for example, Su and others, 2004; Burow and others, 2005) greatly extends the temporal range and the spatial scale over which temperature gradients have been analyzed to use heat as a natural tracer of ground-water movement near streams. This chapter reviews early work that addresses heat as a tracer in hydrological investigations of the near-surface environment, that describes recent advances in the field, and that presents selected new results designed to identify the broad application of heat as a tracer to investigate stream/ground-water exchanges. An overview of field techniques for estimating water fluxes between surface water and ground water is provided here; for a comprehensive discussion with numerous case studies, see Stonestrom and Constantz (2003).

Heat Transfer During Stream/Ground-Water Exchanges

When water is present in a stream channel, heat and water transfer because of vapor movement in the streambed sediments generally is negligible relative to heat and water transfer because of liquid water movement. This eliminates the need to address the complex processes of nonisothermal vapor dynamics in porous material when describing heat and water movement below streams. Within the streambed, heat is transferred into and through sediments as a result of three heat-transfer mechanisms. Radiative heat transfer occurs as solar radiation is adsorbed by the streambed surface. This is the dominant mechanism for a dry streambed, but is usually a small component of heat transfer for the streambed beneath a flowing stream. Heat conduction occurs as diffusive molecular transfer of thermal energy between the streambed surface and the underlying sediments. Heat convection and advection often are used interchangeably in hydrology to indicate heat

[1]U.S. Geological Survey, Menlo Park, CA 94025.

[2]U.S. Geological Survey, Carson City, NV 89706.

[3]Now with Philip Williams and Associates, San Francisco, CA 94108.

transfer resulting from the movement of water (or air). For the present work, it is advantageous to partition their definitions as follows. Heat convection is defined as heat transfer occurring because of the movement of water (or air) above a streambed of dissimilar temperature. Heat advection is defined as the heat transfer that occurs during the movement of water (or air) through the streambed. This alternative definition is useful for the application of heat as a tracer in examining stream/ground water interaction because it aids in delineating between heat transfers as a result of ground-water movement (advection) in contrast to surface-water movement (convection). Thus, heat conduction, convection, and advection all contribute to heat transfer across the stream/streambed boundary, but determination of heat advection is the focus in examining stream/ground-water interaction.

Commonly, all three heat-transfer mechanisms occur simultaneously within stream environments. For example, all three mechanisms occur in a losing stream reach as water infiltrates into the streambed, then percolates through the sediments, potentially recharging the water table. Convective heat transfer occurs between the stream and sediments as stream water flows over the sediments. As a result of this convection transfer, conductive heat exchange occurs between the surface sediments and sediments at depth. Simultaneously, advective heat transport occurs as water infiltrates into the sediment and percolates in a downward (but usually not vertical) direction. The daily and (or) annual temperature extremes are attenuated and delayed with depth in the streambed sediments. The attenuation of temperature extremes is determined by the bulk volumetric heat capacity of the sediments as heat is rapidly exchanged at the pore scale. The delay or lag in temperature extremes is controlled by the rate of downward heat transfer, which is dependent on the thermal conductivity of the sediments and the pore-water velocity through the sediments. The greater the heat transfer, the greater the depth of penetration and the shorter the time lag of temperature extremes. In the vicinity of streams, heat usually is transported more rapidly by moving water than through molecular diffusion and, as a result, higher streambed infiltration rates result in the deeper penetration and shorter lags in temperature extremes (for example, Lapham, 1989; Silliman and others, 1995). For a neutral stream reach (one neither gaining nor losing flow), the streambed-temperature gradients are created by convective heat transfer from the stream to the streambed surface, and heat transport into the sediment is determined by heat conduction alone. (Thus, if the Fourier equation for conductive heat transfer can explain the temperature patterns within a streambed, there is no stream/ground-water exchange.) For gaining stream reaches, as was the case for losing and neutral streams, a temperature gradient is created at the streambed surface because of convective heat transfer. As ground water discharges to the stream, however, the stream-temperature extremes are attenuated at shallow depths because of the heat capacity of the discharging ground water, such that the greater the ground-water discharge the greater the attenuation of temperature extremes and the greater the lag in temperature extremes in the sediments (for example, Silliman and Booth, 1993).

These heat-transfer processes also have important ramifications on stream-temperature patterns. Constantz (1998) examined streamflow and stream-temperature patterns on the Truckee River, California, and its tributaries to demonstrate the use of stream-temperature analysis to determine spatial and temporal patterns of exchange in selected reaches. For example, results in this work showed that stream-temperature patterns could be used to demonstrate that the main-stem Truckee River received significant water from bank storage in response to upstream dam releases, whereas the tributary directly downstream from the dam possessed inadequate bank storage to influence post-release stream-temperature patterns (see Constantz, 1998, fig. 10).

Quantitative Analyses of Heat as a Ground-Water Tracer Near a Stream

Rorabaugh (1954) examined correlations between stream temperature and seepage patterns and proposed the measurement of temperature to quantify heat flow, and thus determine streambed seepage indirectly. He indicated that a ground-water model capable of quantifying heat and water fluxes appeared to be the appropriate tool. A physically based, quantitative analysis of heat and water transport through porous materials was introduced by Philip and de Vries (1957). Their analysis resulted in a comprehensive mathematical description of the coupled process of liquid and vapor water transport simultaneous with the transfer of heat in the solid, liquid, and vapor phases of unsaturated porous material. Application of their analysis has demonstrated that the transport of heat and water in the vapor phase often is important in unsaturated soils, and generally dominates in dry environments (for example, Scanlon and Milly, 1994). As the degree of water saturation increases in sediments, heat transport in the vapor phase abruptly declines as the gas phase becomes discontinuous and then vanishes as sediments approach saturation (for example, Stonestrom and Rubin, 1989). As a result, the comprehensive approach developed by Philip and de Vries (1957) is unnecessary for analysis of heat and water fluxes in material that is sufficiently saturated to inhibit macroscopic gas flow. Streambed sediments beneath wetted channels are sufficiently saturated to ignore macroscopic vapor transport.

Suzuki (1960) and Stallman (1963, 1965) were able to use a single-phase approach to predict water fluxes through saturated sediments, based on measured ground-water temperatures. Their work formed the basis for examination of flow in environments ranging from deep ground-water systems (Bredehoeft and Popadopulos, 1965) to humid hillslopes (Cartwright, 1974). Stallman (1963) presented a general equation describing the simultaneous flow of heat and fluid in the earth. He indicated that ground-water temperatures could be used to determine the direction and rate of water movement. He also indicated that temperatures in combination with hydraulic gradients could be used to estimate sediment hydraulic conductivity. Stallman's equation for the simultaneous transfer of heat and water through saturated sediments for the one-dimensional case of vertical flow (z direction) is as follows:

$$K_t \frac{\partial^2 T}{\partial z^2} - qC_w \frac{\partial T}{\partial z} = C_s \frac{\partial T}{\partial t}, \quad (1)$$

where

K_t is the thermal conductivity of the bulk streambed sediments in W/(m °C);

T is temperature in degrees Celsius;

q is the liquid water flux through the sediments in meters per second;

C_w and C_s are the volumetric heat capacity of water and the bulk sediment in J/(m³ °C), respectively;

and

t is time in seconds.

The value of q is controlled by the Darcy equation as the product of the hydraulic conductivity, K, and the total head gradient, h. When q is zero, the equation reduces to the Fourier equation for the transfer of heat by conduction, and when q is large, advection dominates the transfer of heat, as well as the change of temperature throughout the porous material.

Thermal parameters can be estimated, given some knowledge of streambed materials. The heat capacity of the sediments can be estimated by the following:

$$C_s = f_s(c_s \rho_s) + f_w(c_w \rho_w) + f_a(c_a \rho_a). \quad (2)$$

where f_s, f_w, and f_a are the volumetric fractions of the sediment, water, and air, respectively; c_s, c_w, and c_a are specific heats in J/(kg °C) of the sediment water and air, respectively; and ρ_s, ρ_w, and ρ_a are the densities in kg/m³ of the sediment, water, and air, respectively. The product of the specific heat capacity and the density is the volumetric heat capacity, which is in the range of 0.8×10^6, 4.2×10^6, and 0.001×10^6 J/(m³ °C) for sediments, water and air, respectively (de Vries, 1963).

An alternative approach to describe simultaneous heat and water transport through sediments has been to use an energy transport approach via the convective-dispersion equation (Kipp, 1987). These coupled heat and water-flow equations are included here as equations 3, 4, and 5.

$$\partial \frac{[\theta C_w + (1-\phi)C_s]T}{\partial t} = \nabla \cdot Kt(\theta)\nabla T + \nabla \cdot \theta C_w D_h \nabla T \quad (3)$$
$$ - \nabla \cdot \theta C_w Tq + QC_w T.$$

where

θ is percent volumetric water content;

ϕ is sediment porosity, dimensionless;

D_h is thermomechanical dispersion tensor, in square meters per second;

q is the water flux, in meters per second;

and

Q is rate of fluid source, in seconds.

The left side of the equation represents the change in energy stored in a volume over time. The first term on the right side describes the energy transport by heat conduction. The second term on the right side accounts for thermomechanical dispersion. The third term on the right side represents advective heat transport, and the final term on the right side represents heat sources or sinks to mass movement into or out of the volume. The familiar water-flow equation is as follows:

$$C(\psi, x) \frac{\partial h(x,t)}{\partial t} = \nabla [k(\psi, x) \cdot \nabla h(x,t)]. \quad (4)$$

where

$C(\psi, x)$ is specific moisture capacity, which is the slope of the water-retention curve;

ψ is the water pressure, in meters;

h is the total head, in meters;

x is length, in meters;

and

t is time in seconds (Buckingham, 1907; Richards, 1931).

The thermomechanical dispersion tensor is defined as (Healy, 1990):

$$D_h = \alpha_T |v|\delta_{ij} + \frac{(\alpha_l - \alpha_t) v_i v_j}{|v|}, \quad (5)$$

where α_l and α_t are longitudinal and transverse dispersivities, respectively, in m; δ_{ij} is the Kronecker delta function; v_i and v_j are the ith and jth component of the velocity vector, respectively, in meters per second.

Sediment thermal conductivity, K_t varies with texture and degree of saturation; for the typical case of saturated sediment in a general textural class, however, the uncertainty is greatly reduced. For example, the streambed K_t for a sand channel is likely to range only from 1.0 to 2.0 W/(m °C), so that the value of K_t can be estimated as 1.5 W/(m °C) ± 0.5 W/(m °C) (van Duin, 1963).

After the thermal parameters are assigned, q is estimated via an appropriate heat and mass-transport simulation model (discussed in detail below). Generally, hydraulic conductivity cannot be estimated using this procedure. As opposed to K_t, hydraulic conductivity can vary over several orders of magnitude. Even for saturated conditions, the hydraulic conductivity of sand-textured material can vary from values of 10^{-2} down to 10^{-4} m/s (Freeze and Cherry, p. 29, 1979). For a given sand-textured material, as saturation decreased, values of hydraulic conductivity measured in the laboratory ranged from 10^{-5} meters per second down to 10^{-10} (for example, Constantz, 1982). Consequently, hydraulic conductivity is not isolated from q without an accurate measurement of the hydraulic gradient. For many studies, the goal is to develop estimates of q, so that temperature measurements applied to equation 1 or equation 2 have proved useful in determining the rate of water movement through a region of interest. In some studies (as discussed below), hydraulic gradients are determined on the basis of piezometer measurements, so that values of hydraulic conductivity also are estimated from sediment-temperature patterns.

Using reasonable boundary conditions and thermal and hydraulic parameters, a heat- and water-transport simulation code is run to predict temperature patterns in stream sediments. For the present application, predicted temperature patterns are matched to measured data using an inverse-modeling approach. Specifically, hydraulic information, such as stream stage, are determined, temperatures are monitored in the stream and streambed, and predicted temperatures then are compared with measured temperatures by using trial-and-error methods or a parameter-estimation code.

Temperature Instrumentation

Background

Measurement of temperature gradients in the sediments is required to estimate the rate of heat transfer through the streambed. Measurement of temperature over time at two or more depths within the stream/ground-water system is the minimum temperature data needed to estimate heat and water fluxes in the domain bounded by the temperature measurements. When it is desirable to separate water fluxes into hydraulic conductivity and the hydraulic-gradient components, measurements of hydraulic gradients are required in addition to temperature gradients. Accuracy in estimating thermal parameters sometimes is improved through laboratory analysis of sediment samples, especially for variables in equation 2 and equation 4; however, the spatial variability of textures in fluvial environments often diminishes the effectiveness of coring efforts, such that an estimate based on the bulk textural class over the domain of interest may be a more prudent approach. Operationally, measurements of temperature in the stream environment involves logistical issues, which generally do not occur in forest or agriculture settings (for example, Jaynes, 1990). In the stream environment, fluvial processes create installation challenges that often have to be overcome on a site-by-site basis. Some streams are wide and shallow with a mantle of boulders, whereas other streams are deep with steep banks. Furthermore, damage to or loss of temperature equipment, because of high streamflows, is an issue unique to streams. Equipment selection and installation methods are usually site specific, though two common requirements are equipment that is sufficiently durable in high flows and an installation procedure that avoids preferential flow of pore water along the length of the equipment embedded in the streambed. Often, the manner in which temperature is measured may differ for ephemeral channels as compared with perennial channels.

Figure 1 provides a qualitative, pictorial description of the thermal and hydraulic responses to the four possible states of a streambed—a gaining stream, a losing stream, a dry ephemeral channel, and an ephemeral channel with water. The purpose of this figure is to provide graphical depictions of conditions relevant to the installation of monitoring equipment.

Within each panel of the figure, a hydrograph is depicted on the right, while a pair of thermographs, representing the diurnal pattern in the stream and streambed temperatures, are depicted on the left. For the case of a gaining stream (fig. 1A), the hydraulic gradient is upward as indicated by the positive water pressure in the observation well relative to the stream stage. The stream is shown with a large diurnal variation in water temperature, but the sediment temperature has only a slight diurnal variation. The diurnal variation in the sediment is due to the inflow of ground water to the stream, which is generally of constant temperature on the diurnal time scale. Any variation in sediment temperature is a result of a change in the balance between downward conductive transport of heat and upward advective transport of heat. Thus, for a high inflow of ground water, the sediment temperature will have no diurnal variations, whereas for a slight inflow of ground water, the sediment will have a small diurnal variation in temperature (which will be increasingly damped with depth). Consequently, shallow installation of temperature equipment (in the observation well or directly in the streambed) is desired for a gaining stream reach in order to detect significant temperature variations. For the case of a losing stream (fig. 1B), the downward hydraulic gradient transports heat from the stream into the sediments. The combined conductive and advective heat transport can result in large diurnal fluctuations in sediment temperature. Furthermore, because ground water is not flowing into the stream, stream-temperature variations generally are larger than those for gaining streams (Constantz, 1998). Consequently, deeper installation of temperature equipment (in the observation well or directly in the streambed) may be in order for losing streams. For a dry streambed (fig. 1C), pore-water pressures are negative relative to atmospheric pressure, and, thus, are not measurable in a observation well. The streambed may have extremely high variations in diurnal temperature because of radiative heat transfer; however, the combined effects of low K values and no advective heat transport results in negligible diurnal variations in sediment temperatures below the shallowest (for example, 10 centimeters) streambed depths. For ephemeral stream channels (fig. 1D), a dynamic temperature pattern exists at the initiation of streamflow. Again, the observation well remains empty because of negative pore-water pressures until mounding of the water table results in water entry into the well. For the ephemeral case, convective, conductive, and advective heat transport all contribute to the rapid responses in the streambed surface and underlying sediments, as seen in the abrupt response of the streambed thermograph.

Direct Versus Indirect Measurements

Water and sediment temperatures can be measured directly by inserting a temperature probe (that is, thermistor wire, thermocouple wire, or platinum resistance thermometer wire) into the medium of interest, or indirectly by inserting the probe to a depth of interest in an observation well. In either case, the selected temperature probe is connected to a data logger. Within

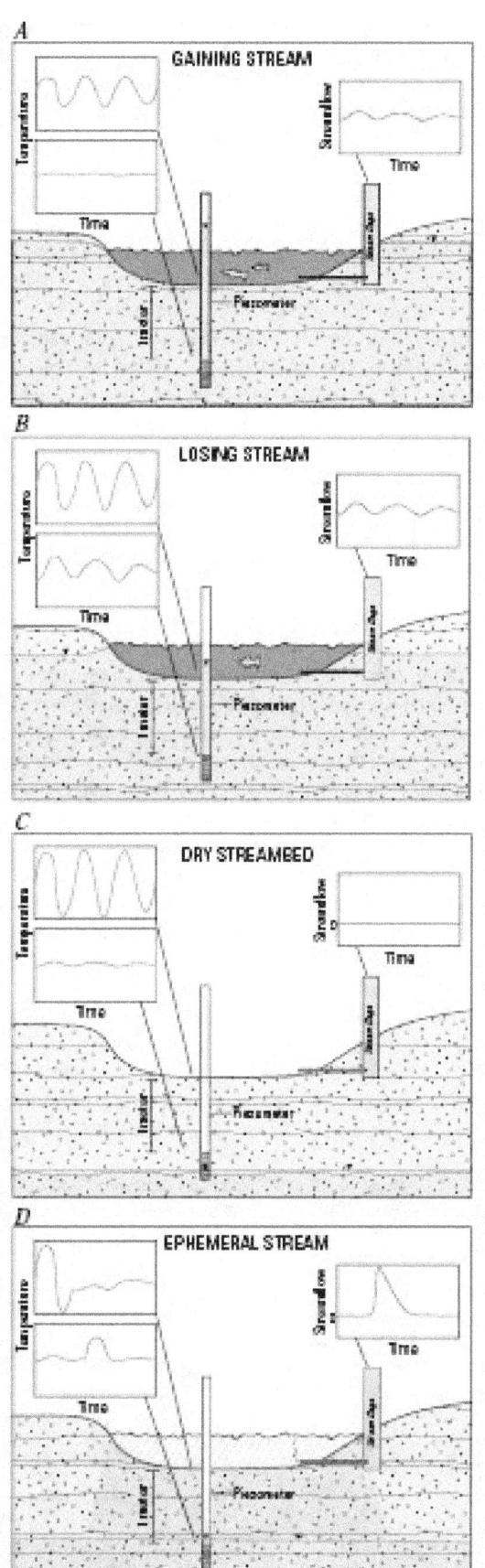

Figure 1.   A qualitative description of thermal and hydraulic responses to four possible states of a streambed: *A*, gaining stream, *B*, losing stream, *C*, dry streambed, and *D*, ephemeral stream. Thermographs and hydrographs are displayed in the upper left and right corners, respectively.

observation wells, temperature can be monitored with temperature-logging equipment on a specific schedule such as hourly, daily, or monthly (see Lapham, 1988), or alternatively, temperature can be continuously monitored at fixed locations within the observation well, using either a series of temperature probes at specific depths, or a series of single-channel, submersible microdata loggers tethered at several locations in the observation well (see Bartolino and Niswonger, 1999). There has been some concern about heat conduction down the observation well, and Lapham (1988) suggested that at least the first meter below the streambed may be influenced by the upper boundary. Another concern has been the development of a convection cell in observation wells that would redistribute heat within the well. For a typical shallow observation well with a 0.05-meter-diameter opening, Samuels (1968) calculated that a convection cell 0.5 meter in length could be established at the top of the water column during periods of large upward thermal gradients (for example, during winter). Samuels demonstrated that for large upward thermal gradients, the temperature in the upper water column could be erroneously high by 0.5 °C. These two complications indicate that measurements in the shallow water column within a observation well may less accurately represent the temperature in the surrounding sediments than values obtained deeper in the well-water column. These complications will be addressed in the final section of this chapter.

Observation wells generally are air filled in ephemeral stream channels, resulting in poor estimates of diurnal temperature variations due to the extremely low thermal conductivity of the air cavity. Thus, streambed temperatures need to be taken directly in the sediment rather than in the observation wells for ephemeral channels. Vertical arrays of thermocouples or thermistors (or temperature nests) have been successfully installed directly into flowing ephemeral channels (Thomas and others, 2000). Thermistor or thermocouple wires are inserted down a drill hole, and as the drill stem is withdrawn, saturated sediment collapses into the hole, thus inhibiting preferential flow. Wires then are run horizontally, either bare or in conduit, to a data-acquisition system. For dry ephemeral channels, thermistor or thermocouple wires are installed into the streambed in a similar fashion as for the flowing case; however, backfilling with either native materials or with diatomaceous earth is necessary to inhibit preferential flow next to the wires. Instrumentation also can be installed horizontally from a trench excavated along the bank (see Ronan and others, 1998). Recently, single-channel, submersible microdata loggers have been successfully installed in dry streambeds using a drill rig and backfill (Bailey and others, 2000). For a detailed description of numerous options for using and monitoring of sediment temperature, see Stonestrom and Constantz (2003).

Methods to Analyze Streambed Temperatures

Several researchers have developed simplifying assumptions for specific hydrological conditions that preclude the necessity of using a heat- and ground-water-transport simulation model. A simplistic, first-approximation approach was developed for the case in which pore-water velocities are sufficiently high such that heat transport by conduction is negligible compared with heat transport by advection. This case is typical during flow events in many ephemeral streambeds (see fig. 1D), and common in perennial stream channels where a dense clay layer is absent. For those cases in which conduction is small compared to advection, pore-water velocity, v, is approximated by:

$$v = V_T \frac{1}{\theta} \frac{C_S}{C_W} . \qquad (6)$$

where

V_T is the vertical velocity of the temperature peak (the "wave" or "front") down into the streambed sediments.

This simplification has been shown to work well in a laboratory column (Taniguchi and Sharma, 1990) and in artificial recharge basin studies (Cartwright, 1974). The flux, q, can be determined from the product of v and θ. The value of θ can be approximated by the porosity of the streambed sediments, although Constantz and others (1988) determined that a value of 0.9θ may be more typical for the initial stages of ponded infiltration. Constantz and Thomas (1996, 1997) have successfully applied this simplification at Tijeras Arroyo, New Mexico, by monitoring temperatures between the surface and a depth of about 3 meters during ephemeral stream-flow events. Stewart (2003) examined the error in using this simplistic approach compared to a complete description of conductive and convective heat transport. Stewart reported that the use of equation 6 could overestimate water fluxes by 30 percent for cases in which heat conduction is a significant component of the total heat flux within streambeds.

Silliman and others (Silliman and Booth, 1993; Silliman and others, 1995) used time-series analysis of stream and sediment temperature patterns in Indiana to identify losing reaches. In a similar fashion to Suzuki (1960) working in rice fields, Silliman and others used a one-dimensional solution to equation 1, with an assumed sinusoidal temperature pattern for upper boundary condition. Silliman and Booth (1995) examined the range of the Peclet number (a measure of advective to conductive transport) for which a solution should be applicable (see Silliman and Booth, 1995, p. 106, for the specific values for Peclet parameters that they chose for a streambed

environment). They concluded that for Peclet numbers of less than 2×10^{-4}, which represent a flux of 8×10^{-8} meters per second, the advective component of the solution is negligible. Thus, this approach may not be useful for the very low water fluxes typical of streambed environments with extensive clay-textured streambeds and (or) very low hydraulic gradients.

Ronan and others (1998) used the heat- and water-transport simulation code VS2DH (Healy and Ronan, 1996) to model the ground-water-flow pattern below Vicee Canyon, Nevada. The ephemeral stream channel within Vicee Canyon meanders over an alluvial fan on the east side of the Carson Range of western Nevada. Along the fan, temperature was monitored in the stream channel and streambed using a 3-meter by 3-meter grid of 24 thermocouples at three locations. The two-dimensional simulation code was used in an inverse modeling approach to match simulated temperature against measured temperature to estimate heat and water fluxes into or out of the streambed vertically and horizontally. After calibration of the model during one season, simulation results were able to predict streamflow loss and streambed infiltration based only on temperature data. Their results used values for dispersivity of about 0.01 meter, indicating that thermal dispersion does not appear to be significant in this type of environment for this length scale (3 meters). Incorporating two-dimensional temperature patterns as input into the model was useful in demonstrating the asymmetrical pattern of substream ground-water flow as a result of down-canyon ground-water flow as the stream meandered across the fan.

The use of heat as a tracer to examine stream/ground-water exchanges has not been limited to shallow investigations. Deeper monitoring of substream temperatures has been done by Lapham (1989) and Bartolino and Niswonger (1999) to estimate annual patterns of stream/ground-water exchanges, where temperatures were periodically logged in observation wells as deep as 50 meters below the streambed. Long-term temperature monitoring provides a series of temperature profiles that can be useful in characterizing streambed fluxes. Figure 2 shows hypothetical streambed-temperature profiles for a losing stream (downward water flux) compared with a gaining stream (upward water flux) over either a year or a day. The temperature profiles for the annual (or daily) extremes

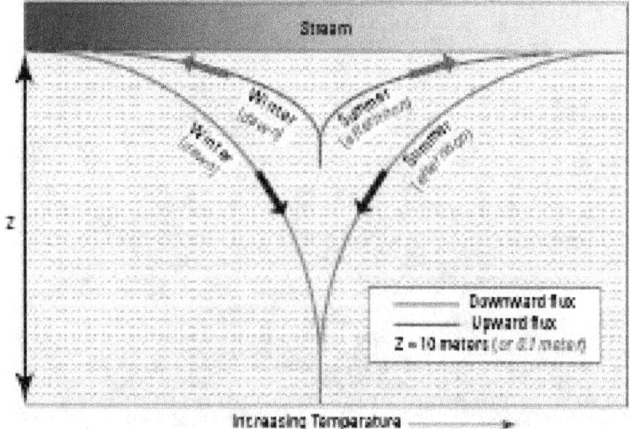

Annual (or Diurnal) Streambed-Temperature Profile

Figure 2.   The streambed temperature profiles (temperature envelopes) for a losing stream (downward water flux) compared to a gaining stream (upward water flux) for the annual example and the diurnal example.

Figure 3. The simulated temperature profile below the Rio Grande in central New Mexico for an optimized value of K (hydraulic conductivity), compared to a value of K one order of magnitude greater or less than the optimal value, and the temperature profile with an optimal value of K, but with an upward value of H (heat) to simulate a gaining stream, based on measured results in Bartolino and Niswonger (1999).

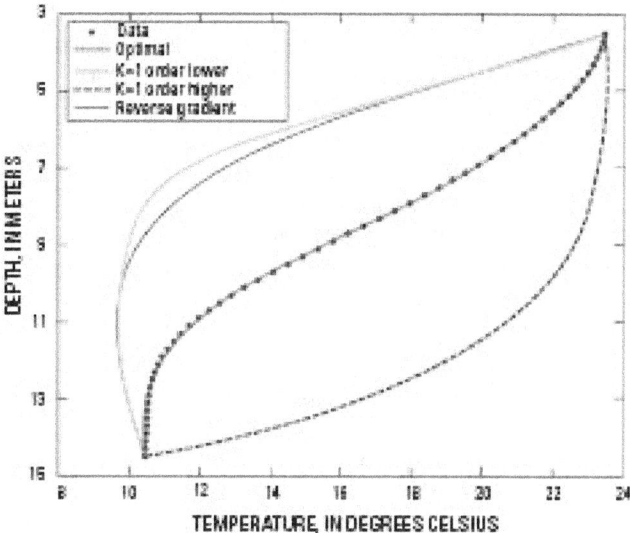

forms a "temperature envelope" for a particular site, within which all other temperature profiles reside. On an annual scale, the January and July temperature profiles typically form an envelope in which other monthly temperature profiles reside. When ground water is discharging to the stream, the annual envelope is collapsed toward the streambed surface, and when the stream is rapidly losing water to the sediments, the envelope extends to great depths. This is true on a daily time scale as well, with the dawn and afternoon temperature profiles forming the daily envelope, in which all other hourly temperature profiles reside. A salient difference between the annual and daily temperature envelopes is the depth scale. See Lapham (1989) for a series of annual and daily example temperature envelopes from streams in the eastern United States.

Figure 3 depicts the effect of changing values of hydraulic conductivity on temperature profiles, based on results for the Rio Grande at Albuquerque, New Mexico (Bartolino and Niswonger, 1999). The figure compares the optimal fit value for hydraulic conductivity (6.7×10^{-5} meters per second) with an order of magnitude increase in hydraulic conductivity, an order of magnitude decrease in hydraulic conductivity, and the optimal hydraulic conductivity with a reversed hydraulic gradient. The large sensitivity of streambed temperature to different hydraulic conditions is clearly apparent. To a lesser extent, simulated flux estimates also are sensitive to uncertainty in thermal parameters. Niswonger and Rupp (2000) used Monte Carlo analysis to examine the relative importance of errors in estimating temperature, K, and C to the resulting simulated water fluxes for Trout Creek, Nevada. When isolating thermal properties, they determined that for the expected mean and standard deviation in thermal parameters, resulting VS2DH simulated water fluxes were most sensitive to uncertainties in sediment temperature and least sensitive to uncertainties in K. In general, predicted fluxes were highly sensitive to variations in hydraulic properties and slightly sensitive to variations in thermal properties for the range of properties reported for sediments.

Example Sites

Study results from two sites are summarized below to provide example applications of the use of heat as a natural tracer of ground-water movement near streams. These sites were chosen because: (1) they are characterized by distinctly different seasonal streamflow patterns; and (2) at the first site, a direct temperature monitoring technique was used, whereas at the second site, an indirect monitoring technique was used.

Both sites were losing stream reaches during the study period; however, the techniques described work equally as well on gaining stream reaches. For an example of direct and indirect temperature measurements used in gaining reaches to estimate upward water fluxes, see Silliman and Booth (1993) for direct temperature measurements made in sediments, and Lapham (1988) for indirect temperature measurements made in observation wells.

Bear Canyon, New Mexico

Bear Canyon is on the eastern edge of Albuquerque, New Mexico. The small ephemeral stream within the canyon is a representative example of more than 100 similar streams that drain from the western flanks of the Sandia and Manzano Mountains into the Middle Rio Grande Basin. The flows in these ephemeral streams have the common characteristics of being bedrock-controlled in their upper gaining reaches, and alluvium-controlled in their lower losing reaches. The streamflow and stream-loss patterns of these stream channels are poorly documented, but their cumulative streambed infiltration might contribute significantly to potential recharge to the basin.

The use of streambed-temperature data was included in a suite of monitoring methods and field-reconnaissance procedures intended to estimate streamflow loss and potential recharge along a reach of Bear Canyon. This reach extends from the exposed bedrock at the mountain-front downslope in a westward direction for about 3 kilometers, at which point the stream channel has been modified as a result of urbanization. The stream is perennial east of the bedrock exposure at the mountain front, and flows rarely extend more than 1 kilometer from the mountain front, though summer monsoons occasionally induce streamflow to the confluence with the Rio Grande, approximately 20 kilometers to the west of the mountain front.

Two temperature-monitoring methods were used within the stream channel of Bear Canyon from 1996 through 1999. Streambed surface temperatures were monitored at sites between the mountain front and the modified reach of the channel, 3 kilometers to the west of the mountain front. Surface-temperature patterns were analyzed as part of the characterization of the spatial and temporal pattern of streamflow in Bear Canyon. Procedures and results for the surface-temperature measurements are described in detail in Constantz and others (2001). Vertical temperature patterns were monitored by using a series of thermocouple wires installed at depths between the streambed surface and about 3 meters below the channel to create a temperature nest in a fashion similar to that described in Thomas and others (2000). Temperature nests were installed at two locations in the middle reach of the Bear Canyon study site, where ephemeral streamflows were expected to be present for extended periods. After backfilling installation holes, the completed temperature nests monitored temperature at 0.40, 0.60, 1.10, 2.10, and 3.10 meters below the streambed surface. Temperatures were monitored at 15-minute intervals via a data logger in an enclosure near the stream channel until September 1999.

Seasonal snowmelt resulted in a gradual progression of the downstream limit of flow down-channel over several months in the spring, followed by a retreat up-channel in early summer. Flashy, summer monsoon streamflow occurred in some, but not all years. Details of the late stages of spring streamflow at one temperature nest in the channel are shown in figure 4. As expected, the greatest diurnal temperature variations are those at a depth of 0.40 meter, and the smallest diurnal temperature variations are those at a depth of 3.10 meters. The abrupt retreat of streamflow upstream in Bear Canyon also is clearly detectable. As streamflow retreated up-channel from this site on June 5, 1999, the abrupt transition from advection-dominated heat transport to conduction-dominated heat transport is quite distinct. Reduced magnitudes in diurnal

variations in streambed temperature result from the loss of advective heat transport with the cessation of streamflow into the streambed.

The streambed-temperature profiles generated at temperature nests during annual spring streamflow were used as input in a fitting procedure that compared measured temperatures to simulated temperature using VS2DH in order to estimate streambed infiltration rates. A commercially available optimization program was used to determine the streambed-sediment hydraulic conductivity from the best fit between the simulated streambed temperatures and measured sediment temperatures. Figure 5 shows sediment temperatures during June 1997 at four depths below the streambed at a vertical temperature site approximately 275 meters west of the mountain front. The figure also shows the simulated best fit at 0.60 meter using an optimization program. The measured streambed temperatures were applied as the upper thermal boundary condition, and measured hydraulic gradients and stream stage were used for hydraulic-boundary conditions. Saturated conditions existed below the stream channel as determined by measuring water levels with piezometers set in the streambed. An optimized seepage rate of 0.75 meter per day resulted in the fit for a depth of 0.6 meter as shown in figure 5 (fits for 0.4 and 1.1 meter depth were comparable but not shown in the figure for clarity of individual thermographs). Optimized simulations for the duration of spring streamflow for this site indicated an average vertical streambed seepage rate of 0.77 meter per day. The consistency in the estimated seepage rate over the duration of the spring season indicates that neither the hydraulic conditions nor the streambed sediments in Bear Canyon were transient. This magnitude of streambed infiltration persisted until retreat of streamflow up-canyon, at which time water fluxes rapidly declined during drainage of the streambed.

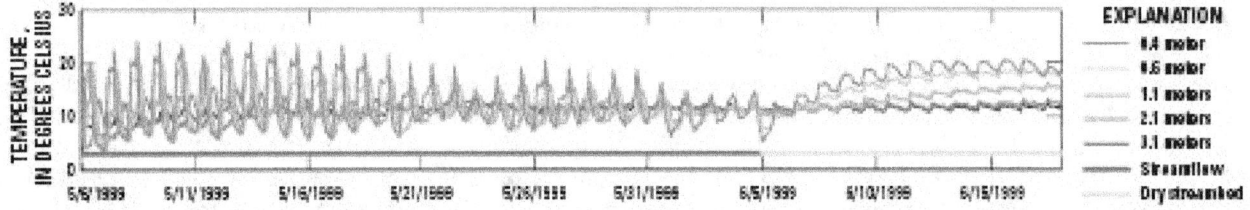

Figure 4.   The streambed-temperature patterns resulting from late stages of continuous spring streamflows progressing to abrupt cessation of streamflow, approximately 275 meters west of the mountain front in Bear Canyon, New Mexico, during June 1999.

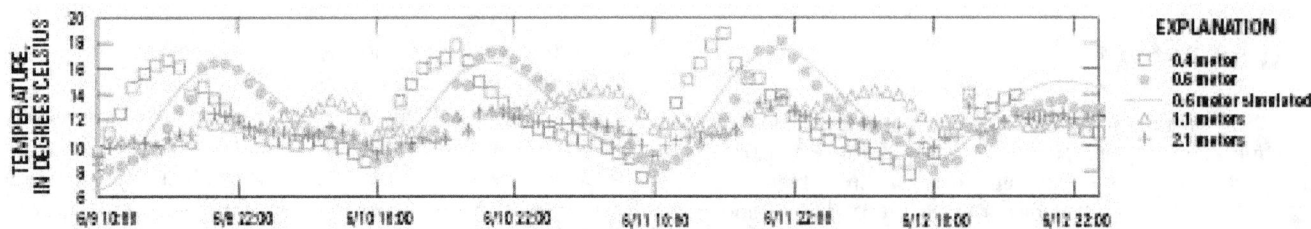

Figure 5.   The streambed-temperature patterns compared to optimized (simulated) sediment temperatures at a depth of 0.6 meter, approximately 275 meters west of the mountain front in Bear Canyon, New Mexico, during June 1997.

Santa Clara River, California

The Santa Clara River is in southern California, flowing from the San Gabriel Mountains approximately 200 kilometers to the Pacific Ocean. In the upper reaches, the gradient is steep, and the stream generally flows over bedrock with a steady gain of ground water. In the middle reaches, the stream flows over a wide sandy channel, resulting in large diurnal stream-temperature fluctuations, as well as substantial potential for stream/ground-water interaction. A 17-kilometer study section was defined in the middle reaches of the river, and a variety of hydrological properties were monitored using a range of surface- and ground-water instrumentation. As part of this larger study, an observation well was installed in the deepest section of a losing reach (referred to as SCR5) in October 1999. The observation well was approximately 4 meters in length with a 0.08-meter internal diameter. The observation well was driven approximately 2.5 meters into the streambed, at which time the drive point was driven from the bottom of the observation well. Temperature between the streambed and the bottom of each observation well was monitored by tethering one single-channel, submersible temperature microdata logger outside the observation well to monitor stream temperature and by tethering three microdata loggers inside the observation well at about 0.6, 1.2, and 2.4 meters below the streambed. The VS2DH simulation code was used to compare one-dimensional simulated temperatures with measured temperatures with a best-fit trial and error match, in order to estimate streambed-percolation rates. Temperature at depth for SCR5 during October 1999 varied during periods in which the stream did not flow and when the stream did flow at this observation-well site. Simulation results matched measured data very well during streamflow, but resulted in a poor match during the intermediate no-flow period at all depths monitored (fig. 6). The poor match during this period is expected because of water drainage from inside the observation well. Thus, microdata loggers suspended in the air-filled interior of the observation well were thermally isolated from the adjacent

sediment during the no-flow period. Once streamflow returned to this location in the stream, the microdata loggers again were submerged and able to effectively monitor sediment temperatures. Consequently, the simulated results probably more correctly matched the sediment temperature during the no-flow period than did the microdata loggers. Direct burial of temperature equipment in the streambed sediments would have avoided the difficulty in monitoring temperature during this period without streamflow. The best-fit simulated temperatures shown in figure 6 resulted in a streambed infiltration rate of 1.8 meters per day, and based on the measured hydraulic gradient in the observation well of 0.41 meter per meter, the derived hydraulic conductivity was 5.1×10^{-5} meters per second.

Temperatures logged beneath the streambed at 0.3 meter varied in comparison to temperature logged inside the piezometer at the same depth as a function of presence or absence of flowing water in the stream (fig. 7). The agreement between the streambed and piezometer temperatures is excellent when streamflow is present and, as expected, agreement is poor during the no-flow period because of drainage of the piezometer and resultant thermal isolation of the microdata logger. The period of excellent agreement between the measured temperatures probably is a result of the strong advective transport of heat at SCR5, such that conduction of heat down the piezometer is small relative to the total transport of heat. Further research is needed to determine the depth of influence of piezometer heat conduction for stream sites where heat conduction is the dominant heat-transport process with the streambed. Realistically, in environments where conduction is the dominant mechanism of heat transport, piezometer design needs to incorporate features that inhibit heat transport vertically along the piezometer while still allowing heat transport horizontally from the sediments to the piezometer. Though the results depicted in figure 7 show good agreement between temperatures, the flux range for good agreement warrants further examination.

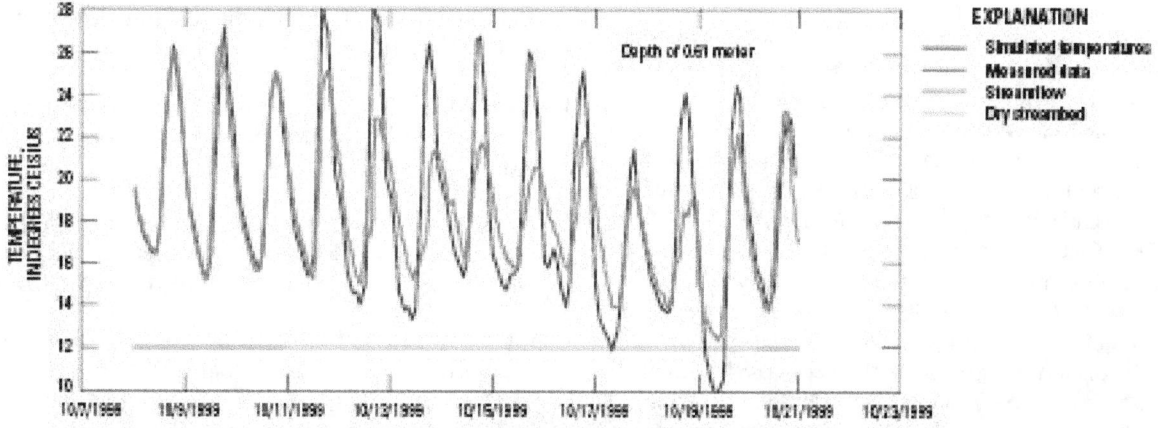

Figure 6.   The measured sediment temperatures compared with simulated streambed temperatures at a depth of 0.6 meter below the streambed surface at observation site SCR5 on the Santa Clara River, California, during October 1999. (Note that streamflow ceased and resumed at the site in the middle of the period of record, as indicated by the horizontal lines at the bottom of the graph.)

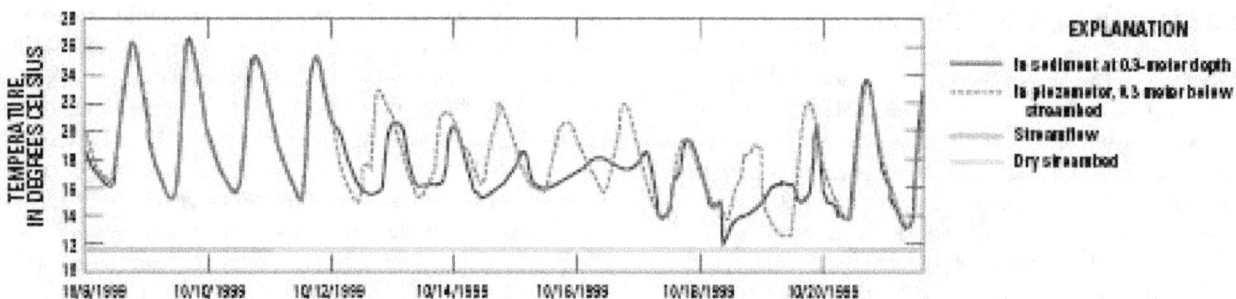

Figure 7. A comparison of streambed temperatures measured directly in the sediments and temperatures measured inside the observation well for observation site SCR5 at a depth of 0.3 meter below the streambed surface on the Santa Clara River, California, during October 1999. (Note that streamflow ceased and resumed at the site in the middle of the period of record.)

Summary

In summary, the measurement and analysis of temperature gradients in streambed sediments provide qualitative patterns and quantitative estimates of rates and direction of water movement through sediments. Both the temporal range, and spatial scale, over which temperature gradients have been analyzed to use heat as a natural tracer of ground-water movement near streams has been greatly extended by numerous recent case studies. Currently, research is ongoing in the areas such as thermal and hydraulic parameter optimization and time-series analysis of temperature gradients to expand the use of heat as a natural tracer in more complex, highly heterogeneous environments.

Recent improvements in acquisition of sediment temperatures and in simulation modeling of heat- and ground-water transport are leading to widespread implementation of methods in which heat is used to trace ground-water fluxes near streams. This chapter provides a brief historical review of the use of heat as a tracer of shallow ground-water movement, and details current theory used to estimate stream/ground-water exchanges. Techniques for installation and monitoring of temperature and stage equipment are discussed in detail for a range of hydrological environments. These techniques are divided into either direct temperature measurements in streams and sediments or indirect measurements in observation wells. Methods of analysis of acquired temperature measurements include analytical solution, heat- and water-transport simulation models, and simple heat-pulse arrival-time procedures. Temperature and derived-flux results are presented for field sites in Bear Canyon, New Mexico, and the Santa Clara River, California. Direct monitoring of temperatures in the sediments below Bear Canyon resulted in estimates of streambed infiltration of 0.75 meter per day, whereas indirect monitoring of sediment temperature using observation wells installed in the Santa Clara River, resulted in streambed-infiltration rates of 1.8 meters per day. The accuracy of measurements within piezometers was confirmed by comparing sediment temperatures acquired directly in sediments with temperatures acquired in a piezometer at the same depth.

References

Bailey, M.A., Ferre, P.A., and Hoffmann, J.P., 2000, Numerical simulation of measured streambed-temperature profiles and soil hydraulic properties to quantify infiltration in an ephemeral stream, American Geophysical Union, 2000 Fall meeting supplement, San Francisco, California, December 15–19, 2000: EOS, Transactions, American Geophysical Union, v. 81, no. 48, p. F501–F502.

Bartolino, J.R., and Niswonger, R.G., 1999, Numerical simulations of vertical ground-water flux of the Rio Grande from ground-water temperature profiles, central New Mexico: U.S. Geological Survey Water-Resources Investigations Report 99–4212, 34 p.

Bouyoucos, G., 1915, Effects of temperature on some of the most important physical processes in soils: East Lansing, Michigan State University, Michigan College of Agriculture Technical Bulletin 24, 63 p.

Bredehoeft, J.D., and Papadopulos, I.S., 1965, Rates of vertical groundwater movement estimated from Earth's thermal profile: Water Resources Research., v. 1, no. 2, p. 325–328.

Buckingham, Edgar, 1907, Studies on the movement of soil moisture: U.S. Department of Agriculture, Bureau of Soils, Bulletin no. 38, 61 p.

Burow, K.R., Constantz, Jim, and Fujii, Roger, 2005, Heat as a tracer to examine dissolved organic carbon flux from a restored wetland: Ground Water v. 43, no. 4, p. 545–556.

Cartwright, K., 1974, Tracing shallow ground water systems by soil temperature: Water Resources Research, v. 10, no. 4, p. 847–855.

Constantz, J.E., 1982, Temperature dependence of unsaturated hydraulic conductivity of two soils: Soil Science Society of America Journal, v. 46, no. 3, p. 466–470.

Constantz, J.E., 1998, Interaction between stream temperature, streamflow, and ground water exchanges in alpine streams: Water Resources Research, v. 34, no. 7, p. 1609–1616.

Constantz, J.E., Herkelrath, W.N., and Murphy, Fred, 1988, Air encapsulation during infiltration: Soil Science Society of America Journal, v. 52, p. 10–16.

Constantz, J.E., and Thomas, C.L., 1996, The use of stream-bed temperatures profiles to estimate depth, duration, and rate of percolation beneath arroyos: Water Resources Research, v. 32, p. 3597–3602.

Constantz, J.E., and Thomas, C.L., 1997, Streambed temperature profiles as indicators of percolation characteristics beneath arroyos in the Middle Rio Grande Basin, USA: Hydrological Processes, v. 11, p. 1621–1634.

Constantz, J.E., Stonestrom, D.A., Stewart, A.E., Niswonger, R.G., and Smith, T.R., 2001, Evaluating streamflow patterns along seasonal and ephemeral channels by monitoring diurnal variations in streambed temperature: Water Resources Research, v. 13, no. 2, p. 317–328.

de Vries, D.A., 1963, Thermal properties of soils, in van Wijk, W.R., ed., Physics of the plant environment: Amsterdam, North-Holland Publishing Company, p. 210–235.

Freeze, R.A., and Cherry, J.A.,1979, Ground water: Englewood Cliff, New Jersey, Prentice-Hall, 604 p.

Healy, R.W., 1990, Simulation of solute transport in variably saturated porous media with supplemental information on modification of the U.S. Geological Survey's computer program VS2D: U.S. Geological Survey Water-Resources Investigations Report 90–4025, 125 p.

Healy, R.W., and Ronan, A.D., 1996, Documentation of computer program VS2DH for simulation of energy transport in variably saturated porous media—Modification of the U.S. Geological Survey's computer program VS2DT: U.S. Geological Survey Water-Resources Investigation Report 96–4230, 36 p.

Jaynes, D.B., 1990, Temperature variations effects on field measured infiltration: Soil Science Society of America Journal, v. 54, no. 2, p. 305–312.

Kipp, K.L., 1987, HST3D—A computer code for simulation of heat and solute transport in three-dimensional ground-water systems: U.S. Geological Survey Water-Resources Investigations Report 86–4095, 517 p.

Lapham, W.W., 1988, Conductive and convective heat transfer near stream: Tucson, University of Arizona, Ph.D. dissertation, 315 p.

Lapham, W.W., 1989, Use of temperature profiles beneath streams to determine rates of vertical ground-water flow and vertical hydraulic conductivity: U.S. Geological Survey Water-Supply Paper 2337, 35 p.

Niswonger, R.G., and Rupp, J.L., 2000, Monte Carlo analysis of streambed seepage rates, in Wigington, P.J., and Beschta, R.C., eds., Riparian ecology and management in multi-land use watersheds: American Water Resources Association, p. 161–166.

Richards, L.A., 1931, Capillary conduction of liquids through porous mediums: Physics, v. 1, p. 318–333.

Ronan, A.D., Prudic, D.E., Thodal, C.E., and Constantz, Jim, 1998, Field study and simulation of diurnal temperature effects on infiltration and variably saturated flow beneath an ephemeral stream: Water Resources Research, v. 34, no. 9, p. 2197–2153.

Phillip, J.R., and deVries, D.A., 1957, Moisture movement in porous materials under temperature gradients: Eos, Transactions, American Geophysical Union, v. 38, no. 2, p. 222–232.

Rorabaugh, M.I., 1954, Streambed percolation in development of water supplies: U.S. Geological Survey Ground Water Notes on Hydraulics, no. 25, 13 p.

Samuels, E.A., 1968, Convective flow and its effect on temperature logging in small-diameter wells: Geophysics, v. 33, no. 6, p. 1004–1012.

Scanlon, B.R., and Milly, P.C.D., 1994, Water and heat fluxes in desert soils 2—Numerical simulations: Water Resources Research, v. 30, p. 721–733.

Silliman, S.E., and Booth, D.F., 1993, Analysis of time-series measurements of sediment temperature for identification of gaining versus losing portions of Judy Creek, Indiana: Journal of Hydrology, v. 146, no. 1–4, p. 131–148.

Silliman, S.E., Ramirez, Jose, and McCabe, R.L., 1995, Quantifying downflow through creek sediments using temperature time series—One-dimensional solution incorporating measured surface temperature: Journal of Hydrology, v. 167 no. 1–4, p. 99–119.

Stallman, R.W., 1963, Methods of collecting and interpreting ground-water data: U.S. Geological Survey Water-Supply Paper 1544-H, p. 36–46.

Stallman, R.W., 1965, Steady one-dimensional fluid flow in a semi-infinite porous medium with sinusoidal surface temperature: Journal of Geophysical Research, v. 70, no. 12, p. 2821–2827.

Stewart, A.E., 2003, Temperature-based estimates of streamflow patterns and seepage losses in ephemeral channels: Stanford, California, Stanford University, Ph.D. dissertation, 248 p.

Stonestrom, D.A., and Constantz, J.E., eds., 2003, Heat as a tool for studying the movement of ground water near streams: U.S. Geological Survey Circular 1260, 96 p.

Stonestrom, D.A., and Rubin, Jacob, 1989, Air permeability and trapped-air content in two soils: Water Resources Research, v. 25, no. 9, p. 1959–1969.

Su, G.W., Japerse, James, Seymour, Donald, and Constantz, J.E., 2004, Estimation of hydraulic conductivity in an alluvial system using temperature: Ground Water, v. 42, no. 6, p. 890–901.

Suzuki, Seitaro, 1960, Percolation measurements based on heat flow through soil with special reference to paddy fields: Journal of Geophysical Research, v. 65, no. 9, p. 2883–2885.

Taniguchi, Makoto, and Sharma, M.L., 1990, Solute and heat transport experiments for estimating recharge: Journal of Hydrology, v. 119, no. 1, p. 57–69.

Thomas, C.L., Stewart, A.E., and Constantz, J.E., 2000, Determination of infiltration and percolation rates along a reach of the Santa Fe River near La Bajada, New Mexico: U.S. Geological Survey Water-Resources Investigations Report 2000–4141, 65 p.

van Duin, R.H.A., 1963, The influence of soil management on the temperature wave near the surface: Wageningen, the Netherlands, Wageningen Institute of Land and Water Management, Technical Bulletin no. 29, 21 p.